Biochar
as a
Renewable-Based Material

*With Applications in Agriculture,
the Environment and Energy*

Biochar
as a
Renewable-Based Material

With Applications in Agriculture, the Environment and Energy

Editors

Joan J. Manyà
University of Zaragoza, Spain

Gabriel Gascó
Technical University of Madrid, Spain

World Scientific

NEW JERSEY · LONDON · SINGAPORE · BEIJING · SHANGHAI · HONG KONG · TAIPEI · CHENNAI · TOKYO

Published by

World Scientific Publishing Europe Ltd.
57 Shelton Street, Covent Garden, London WC2H 9HE
Head office: 5 Toh Tuck Link, Singapore 596224
USA office: 27 Warren Street, Suite 401-402, Hackensack, NJ 07601

Library of Congress Cataloging-in-Publication Data
Names: Manyà, Joan J., editor. | Gascó, Gabriel, editor.
Title: Biochar as a renewable-based material : with applications in agriculture,
    the environment and energy / editors, Joan J. Manyà, University of Zaragoza, Spain,
    Gabriel Gascó, Technical University of Madrid, Spain.
Description: New Jersey : World Scientific, [2020] | Includes bibliographical references and index.
Identifiers: LCCN 2020020783 | ISBN 9781786348968 (hardcover) |
    ISBN 9781786348975 (ebook) | ISBN 9781786348982 (ebook other)
Subjects: LCSH: Biochar--Industrial applications. | Renewable natural resources. | Soil amendments.
Classification: LCC TA455.A63 B56 2020 | DDC 631.8/6--dc23
LC record available at https://lccn.loc.gov/2020020783

British Library Cataloguing-in-Publication Data
A catalogue record for this book is available from the British Library.

For any available supplementary material, please visit
https://www.worldscientific.com/worldscibooks/10.1142/Q0262#t=suppl

Desk Editors: Balamurugan Rajendran/Michael Beale/Shi Ying Koe

Typeset by Stallion Press
Email: enquiries@stallionpress.com

Preface

Biochar is probably one of the topics on which the research community has focused the greatest number of research activities in the last 20 years. The numerous potential benefits of biochar — in a broad range of applications — can explain why a growing number of research studies have been published in several disciplines, including, for instance, soil science, environmental science and engineering, chemical engineering and materials science and engineering.

Despite the large number of available publications, in the form of dedicated books and reviews, further efforts are still needed in connecting and integrating the different scientific and technical disciplines that share the objective of developing biochar technology in several fields of application. With this in mind, the objective of the present book is to provide readers an overall vision of the biochar science and technology, with special emphasis on engineered technologies for biochar production, recent environmental and agronomic applications (e.g. biochar as growing media component or biochar application for mine land reclamation), some emerging biochar applications in different fields (e.g. energy storage and catalysis) and impact of biochar on environmental sustainability.

Special attention is focused on highlighting the need to gain knowledge on the links between feedstock properties, production process and conditions and possible uses of biochar. This becomes essential to guide research and technology towards the production of engineered biochars with the best properties to be used in a given application.

This book, *Biochar as a Renewable-Based Material: With Applications in Agriculture, the Environment and Energy,* is designed as a textbook for graduate and postgraduate courses as well as a handbook for early-stage scientists, policy makers and potential technology customers. The book is written by internationally recognized scientists with complementary backgrounds.

Our sincere thanks to World Scientific, particularly to Dr. Merlin Fox, for bringing us this opportunity. We also thank Jennifer Brough, Michael Beale and Koe Shi Ying, for guiding us through the publication process. Finally, we would like to sincerely acknowledge all the contributors for their valuable work and dedication.

About the Editors

Joan J. Manyà is Associate Professor of Chemical Engineering in the University of Zaragoza (Spain). His research activity is mainly focused on producing biochar and biochar-derived materials to be used in several applications, such as adsorption in gas phase, heterogeneous catalysis and electrochemical energy storage. Dr. Manyà has served as Principal Investigator of several national and international research projects. Currently, he is coordinating an MSCA European Training Network (Project H2020 MSCA-ITN-2016-721991) composed of 14 Ph.D. researchers, and focused on developing advanced biomass-derived carbons for several applications. Other notable engagements include working as a visiting researcher at the Hawaii Natural Energy Institute (University of Hawaii at Manoa, USA) and scientific collaborations with several researchers worldwide, e.g. Dr. Frederik Ronsse (Ghent University, Belgium), Dr. Ondřej Mašek (The University of Edinburgh, UK), Dr. Niklas Hedin (Stockholm University, Sweden), Dr. Vanessa Fierro (CNRS, France), Dr. Jale Yanik (EGE University, Turkey) and Dr. Elsa Weiss-Hortala (IMT Mines Albi, France).

Gabriel Gascó is Professor of Soil Science in the School of Agricultural, Food and Biosystems Engineering at Technical University of Madrid (UPM, Spain) from 2002. Previously, Dr. Gascó was a lecturer in the Catholic University of Ávila and in the Miguel de Cervantes European University. Dr. Gascó has made research stays in KU Leuven University (Belgium), RMIT University (Australia) and the Institute of Soil Science and Plant Cultivation (Poland). His research interests include the use of pyrolysis and hydrothermal treatment as carbon sequestration technology, the use of both pyrochar and hydrochar in the improvement of agricultural soils, the recovery of degraded land and contaminated soils and as growing media components. Dr. Gascó has published more than 65 papers in international journals and has done more than 100 presentations in conferences about these topics.

https://doi.org/10.1142/9781786348975_fmatter

About the Contributors

Eliana Cárdenas-Aguiar works at the National Environmental Licensing Authority (ANLA) of Colombia in the Department of Agrochemicals. She holds an M.Sc. in Environmental Engineering and a Ph.D. in Agroengineering at the Technical University of Madrid (UPM, Spain). Dr. Cárdenas has made a research stay in the University of Edinburgh and worked with the UK biochar research centre (UKBRC). Her research is associated with the use of biochar, hydrochars and organic manures to improve soil properties and in general the use of these materials to amended mining soils.

Tamara L. Church earned her Ph.D. in Organic Chemistry while studying catalytic mechanisms at Cornell University (USA), and was a Postdoctoral Fellow in Uppsala (Sweden) again studying the mechanism of a catalytic system. She then worked in CO_2 capture and heterogeneous catalysis at the University of Sydney (Australia) before joining the research group of Professor Niklas Hedin at Stockholm University. There she performed research on materials including organic–inorganic hybrids, hydrothermal carbons, and microporous organic polymers.

Irmina Ćwieląg-Piasecka is a chemist with scientific interests focused on environmental chemistry with particular emphasis on structural studies of organic matter, using spectroscopic techniques. Currently studying the role of humic substances in organic xenobiotics binding in soil as well as the effect of the biochar addition to soil on pesticides immobilization and metal translocation.

Jean Dalmo de Oliveira Marques is a Professor in the Department of Chemistry, Environment and Food at the Federal Institute of Education, Science and Technology of Amazonas (IFAM) (Manaus, Brazil). He was an Agronomist graduate at the Federal University of Manaus — UFAM, has a Master's degree in Agronomy from the University of São Paulo (USP-ESALq) and a Ph.D. in Ecology from the National Institute of Amazon Research (INPA). His research is focused on the dynamics of soil carbon, as well as on physical and water attributes of the Amazon soils.

Niklas Hedin studied with Professor István Furó at the Royal Institute of Technology (Sweden) and graduated with a Ph.D. in Physical Chemistry. He performed postdoctoral studies with Professor Bradley Chmelka at the University of California at Santa Barbara and with Dr. Sebastian Reyes at ExxonMobil Corporate Research Laboratory, Annandale, New Jersey. He is now a Professor of Materials Chemistry at Stockholm University, researching adsorbents for separation or reduction of carbon dioxide and carbon materials derived by hydrothermal carbonization.

Ondřej Mašek is a Lecturer in Engineering Assessment of Biochar at the University of Edinburgh (UEdin). He has established and lead biomass pyrolysis and biochar production research at the UK Biochar Research Centre (UKBRC) at UEdin since its establishment in 2009. Over the past 5 years his team has developed, built and commissioned a unique set of laboratory and pilot-scale facilities for biomass pyrolysis and gasification, representing a total investment of over £500 k by the University and private sponsors. His interests are focused on development of thermochemical conversion processes for biomass valorization by co-production of solid carbon products and chemicals derived from pyrolysis liquids.

Agnieszka Medyńska-Juraszek is a Soil Scientist with the scientific interest focused on environmental chemistry with particular emphasis on heavy metal pollution and remediation of degraded soils with organic amendments. She is the author and co-author of many research publications on soil, biochar and pollution topics, and has been the leader and the main researcher in a project "Biochar as a growing medium for greenhouse vegetable production". She is currently working on biochar and its role in xenobiotic immobilization in soil as well as effects of biochar on soil properties and nutrient availability for plants.

Ana Méndez is a Professor of Material Science and Metallurgical Engineering in the School of Mining and Energy Engineering at the Technical University of Madrid (UPM, Spain). Dr. Méndez has done research stays at the University of Porto and University of Lisbon (Portugal), Université Catholique de Louvain (Belgium), RMIT Melbourne (Australia) and the Centre de Recherche sur la Matière Divisée CNRS (France). Her research is focused on the preparation and characterization of carbon materials and their environmental applications and also in the metal recovery from wastes by hydrometallurgical processes. Dr. Méndez is author of more than 75 articles in international journals and more than 100 presentations in national and international conferences.

Jorge Paz-Ferreiro got his Ph.D. from Universidad de Santiago de Compostela in 2007 and is a Senior Lecturer in Environmental Engineering at RMIT University (Australia). He has published over 100 articles, according to Scopus, with an h-index of 30. He has supervised to completion two Ph.D. students and two Master's students since 2014. Jorge's interests are broad and include soil remediation, soil pollution, nutrient cycling in soil, soil biological activity and pyrolysis.

Frederik Ronsse is the Principal Investigator in the "Thermochemical Conversion of Biomass" research unit and is responsible for managing a team with expertise in thermochemical conversion of biomass. In addition to research activities, he is also the lecturer of several undergraduate and graduate-level courses at the Bioscience Engineering Faculty of the Ghent University and is guest lecturer at the University Ghent Global Campus in Songdo (Incheon), South-Korea. Course topics include biomass conversion, process engineering and thermodynamics.

Guillermo San Miguel is a Lecturer and Senior Research Fellow (PCD-I3) in the School of Industrial Engineering (ETSII) at Universidad Politécnica de Madrid. He holds a B.Sc. in Chemistry, an M.Sc. in Environmental Impact Assessment from University of Wales and a Ph.D. in Environmental Engineering from Imperial College of London. He received a Ramón & Cajal fellowship and a I-3 award for research excellence in 2007 from the Spanish Ministry of Science. His research interests include Life Cycle Assessment (LCA), environmental, economic and social analysis of products, services and organizations, carbon footprint analysis, renewable energies and waste management.

Anthony Edward Szego graduated from the University of Seville, in Spain, with a B.Sc. in Chemistry, finding interest in the field of catalysis. Following this interest, he went on to obtain a Master's Degree in Catalysis from the Cardiff University. He is currently working in Professor Niklas Hedin's group in the Department of Materials and Environmental Chemistry at Stockholm University with a Marie Curie scholarship within the GreenCarbon European Training Network.

Wenceslau Geraldes Teixeira is a soil physicist working for Ministry of Agriculture at the Brazilian Agricultural Research Corporation (Embrapa) in National Soil Center Research (Embrapa Soils, Rio de Janeiro). He also acts as collaborator Professor of Environmental Modelling at the State University of Rio de Janeiro (UERJ). His research is focused on the characterization of physical properties of tropical soils, specially properties involving water retention and soil–plant–atmosphere water fluxes. He also works with geoarchaeology using geophysical techniques to identify human footprints in soil. Dr. Teixeira has been working in the Amazonia basin with Anthrosols (Amazonian Dark Earths, Sambaquis and Geoglyphs) and biochar for more than 20 years. He is the author or co-author of many articles, book chapters and books and has attended and organized numerous international meetings about biochars in soils.

Frank Verheijen is a Physical Geographer, who received his Ph.D. in soil science in 2006 on research of the on-farm economic value of soil organic matter. He has focused his research on soil organic matter: how to manage it, and how it interacts with the wider environment. He started working on biochars in 2009 when he led a team to write a major European Commission review on biochar application to soils. In his current research position at the Centre for Environmental and Marine Studies of the university of Aveiro (Portugal), he focuses on how soil management can improve ecosystem services, in particular elucidating positive as well as negative effects of biochar on soils, and identifying trade-offs in order to build a sustainable biochar system. Dr. Verheijen has authored over 50 publications and his work has been cited by other researchers over 5,000 times.

Peng Zhang started working with nanoporous materials in 2012, obtained his Ph.D. at Uppsala University (Sweden) in 2016 and continued his research work as a Postdoctoral Fellow at Stockholm University in Professor Niklas Hedin's group until 2018. Currently, he is a young lecturer at Yantai University (China), where his research is focused on the applications of functional polymers, especially in energy storage, and drug delivery systems.

Contents

Preface		v
About the Editors		vii
About the Contributors		ix
Chapter 1	Biochar as a Sustainable Resource to Drive Innovative Green Technologies *Joan J. Manyà and Gabriel Gascó*	1
Chapter 2	Biochar Production via Pyrolysis *Frederik Ronsse, Ondřej Mašek and Joan J. Manyà*	35
Chapter 3	Water Holding Capacity of Biochar and Biochar-Amended Soils *Wenceslau Geraldes Teixeira, Frank Verheijen and Jean Dalmo de Oliveira Marques*	61
Chapter 4	Biochar as a Growing Media Component *Agnieszka Medyńska-Juraszek and Irmina Ćwieląg-Piasecka*	85
Chapter 5	Biochar Application for Mine Land Reclamation: Metal Mining *Gabriel Gascó, Eliana Cárdenas-Aguiar, Jorge Paz-Ferreiro and Ana Méndez*	105

Chapter 6   Biochar-based Carbon Materials for Adsorptive
            Separation and Applications in Catalysis   131
            *Tamara L. Church, Anthony E. Szego and
            Niklas Hedin*

Chapter 7   Biochar-based Carbon Materials for Energy-Storage
            Applications   165
            Niklas Hedin and Peng Zhang

Chapter 8   Environmental Assessment of Biochar Using
            a Life Cycle Approach   183
            Guillermo San Miguel

Index   199

https://doi.org/10.1142/9781786348975_0001

Chapter 1

Biochar as a Sustainable Resource to Drive Innovative Green Technologies

Joan J. Manyà[*,‡] and Gabriel Gascó[†]

Escuela Politécnica Superior, Aragón Institute of Engineering Research (i3A), Universidad de Zaragoza, Huesca, Spain

†*School of Agricultural, Food and Biosystems Engineering, Technical University of Madrid, Madrid, Spain*

‡*joanjoma@unizar.es*

Abstract

An overall vision of the biochar science and technology is given in this chapter. The emerging applications of biochar in electrochemical energy storage and catalysis, as well as its environmental applications within the scope of the circular economy, are described. The use of biochar as carbon sequestration technology, the benefits of its utilization as growing media component or its possible use for the recovery of contaminated soils are treated in this chapter. Special attention is focused on highlighting the need to gain knowledge on the links between feedstock properties, production process and conditions and possible uses of biochar. This becomes essential to guide research and technology toward the production of engineered biochars with the best properties to be used in a given application.

1

1.1 Introduction

There are many definitions about the concept of biochar. Lehmann and Joseph defined biochar as a carbon-rich solid obtained by the thermal decomposition of organic matter under a limited supply of oxygen and at relatively low temperatures [Lehmann and Joseph, 2009]; the International Biochar Initiative defines biochar as a solid material obtained from thermochemical conversion of biomass in an oxygen-limited environment [IBI, 2015] while the European Biochar Certificate [Delinat Institute, 2013] defines biochar as a char produced by pyrolysis for use in agriculture (and other non-thermal applications) in an environmentally sustainable manner. According to this last definition, the solid fraction obtained by thermal treatments as torrefaction or hydrothermal carbonisation (HTC) (solid fraction is named hydrochar) cannot be called biochar [Delinat Institute, 2013] although the solid fraction can have a similar end use in agriculture or for other environmental purposes.

Chars can be prepared from biomass with different origins as agriculture and forestry, gardens, kitchens or canteens or other wastes as biosolids [Méndez *et al.*, 2012; Paz-Ferreiro *et al.*, 2012], paper fiber sludge [Méndez *et al.*, 2009] or different types of manure [Cárdenas-Aguiar *et al.*, 2019; Gascó *et al.*, 2018] with several environmental advantages. For example, the application of biochar to the soil can imply the reduction of emissions of CO_2 and greenhouse gases as methane and nitrous oxide emissions (Table 1.1). Traditionally, the main application of biochar has been its application to agricultural soils to increase crop production. The enhancement of agricultural production is due to the improvement of physical, chemical and biological soil properties. For example, biowaste-derived biochar can increase the soil water holding capacity and improve the structural stability of soil [Méndez *et al.*, 2012], increase both cation exchange capacity (CEC) and pH of soil and nutrient content [Paz-Ferreiro *et al.*, 2012]. Also, there are several studies describing how chars can improve biological properties of soils [Gascó *et al.*, 2016; Paz-Ferreiro *et al.*, 2012, 2016] contributing to improved soil quality. On the other hand, the production of biochar and its storage in soils have been suggested as a technology for abating climate change by sequestering carbon estimation such that the total net emissions of CO_2 could be reduced over the course of a century by 130 Pg CO_2 [Wolf *et al.*, 2010].

Table 1.1:　Environmental advantages of biochar soil application.

Agricultural use

1. Carbon sequestration: Reduction of CO_2 emissions.
2. Reduction of methane and nitrous oxide emissions.
3. Positive effect in chemical soil properties: Increase in CEC, reduction in soil acidity and organic matter content.
4. Positive effect in physical properties: Improve water holding capacity, available water, field capacity and soil structure.
5. Increased microbiological activity.
6. Reduction of the amount of organic and inorganic fertilizers needed.
7. Increased quality and yield of crops.

Other advantages

8. Thermal treatment reduces the risk of presence of pathogens.
9. Adsorption of organic compounds: Reduction of bad smells.
10. Control of metals leaching.
11. Reduction of phosphate and nitrate pollution of streams and groundwater.
12. Good growing media to the plants.
13. Reduction of the volume of biowaste to transport to landfill.

1.2　Biochar and Soil Carbon Sequestration

The potential carbon sequestration in soil after biochar application relates with carbon stability of biochar, which depends on the raw material and pyrolysis conditions, mainly pyrolysis temperature. The conditions of biochar preparation have a great influence on the carbon content, type of organic compounds, whether it is easily oxidized, the organic carbon, volatile matter and fixed carbon that are related with the potential carbon sequestration in soil after biochar application. Bruun *et al.* [2011] observed that the cumulative CO_2 emissions from soil were reduced from 11% to 3% application of biochar from wheat straw prepared at 475°C and 575°C, respectively, according to the reduction of the biochar labile fraction with the increase in pyrolysis temperature.

Méndez *et al.* [2013] estimated that CO_2 emissions from a loamy soil after 10 years can be reduced after the application of biosolid-based biochars (rate: 8%) from 301 (pyrolysis temperature: 400°C) to 932 kg CO_2 ha^{-1} (pyrolysis temperature: 600°C) with respect to the direct application of biosolids. Gascó *et al.* [2016] described, after the application of

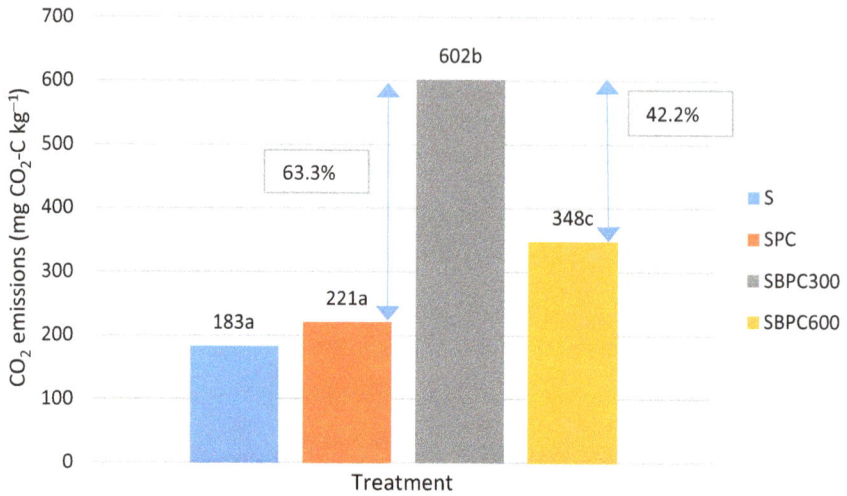

Figure 1.1: C mineralization of soil (S) after application of pig manure (SPC), pig manure biochar prepared at 300°C (SBPC300), pig manure biochar prepared at 500°C (SBPC500) at a rate of 8%. Values followed by the same letter are not significantly different ($P = 0.05$) using the Duncan test.

Source: Adapted from Gascó *et al.* [2016].

biochar prepared from pig manure at 300°C and 500°C, reductions of CO_2 emissions from 42.2% to 63.3% after 219 days and no differences between control soil (classified as Cambisols) and soil amended with the biochar prepared at 500°C (Figure 1.1). Apart of the carbon sequestration, the preparation of chars from organic waste as biosolids or manures, has other environmental benefits, such as the reduction of nitrogen leaching, the recovery of P and K [Wang *et al.*, 2012], reduction of the volume of waste material, improvement of both physical and biochemical soil properties [Liang *et al.*, 2014; Lu *et al.*, 2014] with important implications on agricultural production within the scope of the circular economy due to the valorization of organic wastes.

On the other hand, the pyrolysis may become more expensive due to high moisture content of some raw materials as biosolids, or pig slurry making, in these cases, the HTC more suitable due to it being a thermochemical process of organic matter conversion in the presence of water, under moderate temperatures (180–260°C) and autogenous pressures [Hiltzl *et al.*, 2015], as a result of which HTC requires minimal drying compared with pyrolysis. The produced hydrochar can have a moderate heat value, high carbon content and mesoporous textures [He *et al.*, 2013] with different applications depending on thermal treatment conditions. Gascó *et al.* [2018] concluded that HTC of pig manures could be an interesting method to obtain soil growing media or green roof materials with adequate hydrophysical properties, whereas biochars obtained by pyrolysis of pig manure (Figure 1.2) could be used as soil amendments and that obtained at 600°C can be used for carbon sequestration due to its higher aromaticity.

However, the preparation of hydrochars is also a good strategy for the valorization of organic wastes and for the reduction of CO_2 emissions. Cárdenas-Aguiar *et al.* [2019] observed, during an incubation experiment of 60 days, a reduction of CO_2 emissions between 97.7% and 88.7% for biochars produced at 600°C and 450°C and between 68.8% and 59.0% for hydrochars produced at 240°C and 190°C, with respect to the incubation of a rabbit manure. This increase in carbon sequestration is related with the degree of organic matter humification and thermal stability after both

Figure 1.2:   Pictures of biochars (BPM) and hydrochars (HPM) prepared from pig manure at different temperatures. Biochars: BPM300 (300°C), BPM450 (450°C), BPM600 (600°C). Hydrochars: HPM200 (200°C), HPM220 (220°C), HPM240 (240°C).

Source: Adapted from Álvarez [2019].

pyrolysis and HTC treatments. The difference between the CO_2 emissions between biochars and hydrochars is due to that pyrolysis produces chars with predominantly aromatic structures while HTC results in chars with a higher content in aliphatic structures according to the thermal stability of the materials. The importance of pyrolysis and HTC as carbon fixation technologies for adaptation and mitigation of climate change have been revealed by the roundtable "Biochar: the safe, scalable and shovel-ready carbon sequestration solution" organized by International Biochar Initiative and Husk Ventures in the last UN Climate Change Conference celebrated on December 2019 in Madrid.

1.3 Biochar as Component of Growing Media

More recently, the use of chars as components of growing media has emerged as a new use to reduce the pressure on peatlands, which cover an estimated area of 400 million ha, equivalent to 3% of the Earth's land surface. Peatlands globally represent a major store of soil carbon sink for carbon dioxide and source of atmospheric methane. For example, northern peatlands store around 450 billion metric tons of carbon, which is equivalent to approximately one-third of the global soil carbon stock, which is an important sink for carbon dioxide and source of atmospheric methane [Strack, 2008]. Peatlands are a resource that are not renewable at the human timescale, and nowadays there is a great deal of concern surrounding peatland protection and their overexploitation, which can cause, for example, the autoignition of peatlands (Figure 1.3) and the destruction of a non-renewable resource. For instance, there are necessarily 100 years required for the formation of between 0.8 and 12.1 cm of peat under Spanish weather conditions [Guerrero, 1987].

As growing media component, peat is mixed with components like vermiculite, perlite or clays. Indeed, the price of different growing media ingredients as perlite (50 €/m^3), pumice (40–45 €/m^3), clay (60 €/m^3) or peat (25–40 €/m^3), coupled with the above-mentioned environmental concerns, are pushing towards the use of alternative materials [Álvarez *et al.*, 2017]. Nevertheless, water holding capacity of biochar is lower than other common growing media components, i.e. peat, limiting their ability to fully replace growth substrates in horticulture. However, although biochars do not show adequate chemical and hydrophysical properties for

Figure 1.3:   Autoignition of Zuacorta peatland (Spain) in 1985.
Source: Courtesy of F. Guerrero.

their individual use as growing media, their addition to peat in adequate rates could improve the chemical and physical properties of growing media and, as a consequence, the biomass production of different species [Nieto *et al.*, 2016; Méndez *et al.*, 2015]. Recently, biochar have been tested as growing media component with remarkable results when they are mixed with peat, including increases of yield over 100% [Nieto *et al.*, 2016; Méndez *et al.*, 2015]. Table 1.2 shows some relevant studies about the use of chars as growing media components.

Nevertheless, it is possible to obtain hydrochars from selected biomass by HTC with similar hydrophysical properties as peat. For example, Figure 1.4 shows how the water holding capacity curves at low suction values (until −10 kPa) of two hydrochars prepared from biosolids (HSL) and organic fraction of urban waste (HUW) are very similar to the curve of a commercial brown peat (PT). In this case, the HSL was thermally treated at 205–215°C during 6 h and 17–19 bar, whereas HUW was treated at 190–200°C during 6 h and at a pressure of 17–19 bar. The figure also shows that HSL presents a similar water retention curve to that of PT. Indeed, Álvarez *et al.* [2017] described increments of biomass production of *L. perenne* higher than 180% in growing media prepared with biosolids-derived hydrochar with respect to peat alone.

Table 1.2:   Use of biochar/hydrochar as growing media component.

Feedstock	T (°C)	Crop	Treatments (%, v/v)	Yield (%)	References
Biochar as growing media component					
Organic fraction urban waste	270	Ryegrass	Peat	100	Gascó *et al.*
			Biochar	175	[2018]
			Peat + Biochar (50/50)	274	
Pruning waste	500	Lettuce	Peat	100	Nieto *et al.*
			Peat + Biochar (50/50)	200	[2016]
Deinking sludge	300	Lettuce	Peat	100	Méndez *et al.*
			Biochar	2	[2015]
			Peat + Biochar (50/50)	165	
Crushed wooden boxes	600	Sunflower	Biochar	100	Steiner *et al.*
			Peat	137	[2014]
			Biochar + Peat (25/75)	115	
			Biochar + Peat (50/50)	112	
			Biochar + Peat (75 + 25)	125	
Urban green waste	550	Wheat	Scoria (20% coir) + Biochar (70/30)	121	Cuong *et al.*
			Scoria (20% coir) + Biochar (60/40)	111	[2014]
			Scoria + Biochar (60/40)	111	
Green waste	160–220	Calathea	Biochar (100)	100	Tian *et al.*
			Biochar + Peat (50/50)	163	[2012]
			Peat (100)	133	
Hydrochar as growing media component					
Biosolids	190–200	Ryegrass	Peat	100	Álvarez *et al.*
			Biosolids hydrochar (50/50)	184	[2017]
			Peat + Biosolids hydrochar (50/50)	216	
Organic fraction urban waste	205–215	Ryegrass	Peat	100	Álvarez *et al.*
			Urban waste biochar (50/50)	100	[2017]
			Peat + Urban waste biochar (50/50)	100	

Furthermore, Gascó *et al.* [2018] have prepared hydrochars (range of temperatures: 220–240°C) from pig manures with water holding capacities higher than 66%, this value being between 523% and 595% higher than raw material and more than 200% higher than biochar prepared at 600°C. These authors related the higher water holding capacity with more volume of porous hydrochars in the range between 200 and 30,000 nm, which are related with water content of chars between 33 and 1500 kPa of pressure.

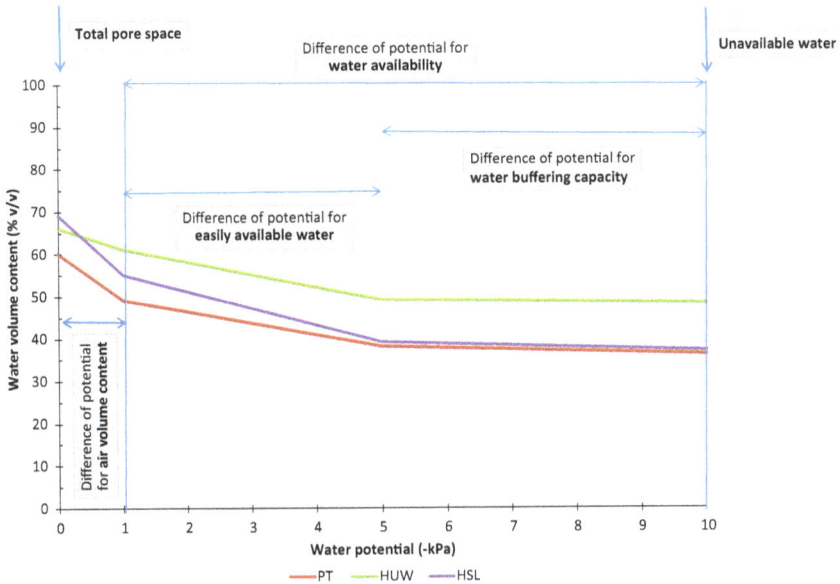

Figure 1.4:   Water retention curves of hydrochars from biosolids (HSL) and organic fraction of urban waste (HUW) and a commercial brown peat (PT).

In summary, it could be possible to prepare hydrochars from selected biomass and adequate HTC conditions, which could have hydrophysical and chemical properties similar to those of peat with applications in the preparation of growing media or other environmental applications as recovery of contaminated or degraded soils.

1.4 Biochar for the Recovery of Contaminated Soils

In the last years, the use of biochar have received attention as soil amendment for the stabilization of organic and inorganic contaminants, with a great amount of studies carried out about this application [Méndez *et al.*, 2009; Sopeña *et al.*, 2012]. With respect to heavy metals, different biochar properties as pH, CEC, presence of oxygenated surface groups, surface area or high content of Ca^{2+}/Mg^{2+} are involved in the immobilization of heavy metals in soils, reducing their mobility and bioavailability [Méndez *et al.*, 2009; Paz-Ferreiro *et al.*, 2014; Uchimiya *et al.*, 2011] due to different processes as precipitation, adsorption and exchange or modification of redox state of the trace metal. This fact together with the improvement of soil pH, CEC, both nutrient and organic matter content, water holding capacity or biological soil properties after biochar addition can be positive to the growth of phytoextractor increasing the metal extraction in multi contaminated soil [Paz-Ferreiro *et al.*, 2014]. This topic will be extensively addressed in Chapter 5, "Biochar Application for Mine Land Reclamation: Metal Mining".

1.5 Emerging Applications of Biochar

Biochar-based materials can also be excellent candidates for applications in adsorption, catalysis and energy storage [Liu *et al.*, 2015]. Despite the fact that raw biochar has poor surface functionality (only some C–O, C=O and OH groups can be present) and very limited surface area (usually < 150 m^2g^{-1}) [Mullen *et al.*, 2010], its porosity and surface chemistry can be relatively easily tuned for a given application, resulting in a very promising platform to synthesize valuable and tailor-made functional materials.

1.5.1 *Biochar activation*

To enhance the performance of biochar-based materials in adsorption or catalytic applications, a controlled porosity and a high surface area are desirable to facilitate high mass transfer fluxes and high active loading [Titirici *et al.*, 2012]. To this end, activated carbons having a hierarchical porosity (comprising macro-, meso- and micropores; as shown in Figure 1.5) can be produced from the raw biochar through either physical or chemical activation. During the physical activation process, the raw biochar is partly gasified with steam or CO_2 at temperatures ranging from 800°C to 1000°C. Physical activation usually leads to carbons having a well-developed microporosity with a very small contribution from mesoporosity [Liu *et al.*, 2015]. During chemical activation, the raw biochar is usually impregnated or mixed with alkali chemicals, such as KOH, an then heated under an inert gas environment at 600–800°C. The mass ratio KOH/biochar appears to be the most influential factor on the porosity of the resulting activated carbon. In this sense, a lower KOH dosage

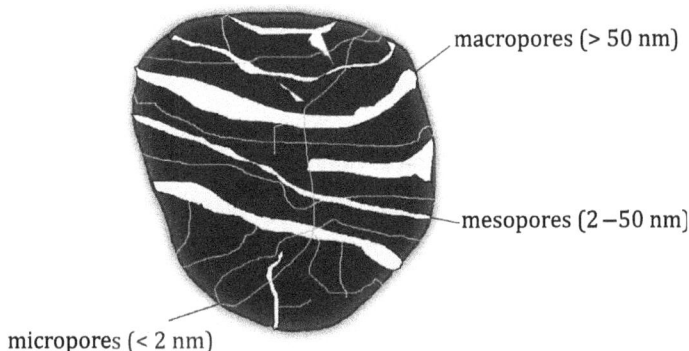

macropores (> 50 nm)

mesopores (2 –50 nm)

micropores (< 2 nm)

Figure 1.5:   Schematic illustration of an activated carbon particle.

(e.g. a mass ratio KOH/biochar lower than 2) can promote the formation of micropores or ultra-micropores (<0.7 nm) with a narrow pore size distribution [Wei *et al.*, 2012]. On the contrary, increasing the KOH/biochar ratio can result in an increase in the specific surface area and total pore volume (with higher contribution from mesoporosity) as a consequence of pore widening [Li *et al.*, 2014].

As an alternative pathway to the two-step production of activated biochars (i.e. carbonization and subsequent activation), single-step production processes have recently received special attention, especially when hydrothermal conversion processes are involved. As an example of one-step process, one can mention the combined HTC and salt templating (e.g. by using eutectic salt mixtures) synthesis process as a very promising pathway to obtain carbon materials with an engineered micro- and meso-porosity [Alatalo *et al.*, 2016]. In addition, it should be highlighted that the introduction of heteroatoms (N,S) can also be conducted via single-step HTC processes. In this respect, S- or N-doped carbons can significantly improve the performance of biomass-derived carbon materials in many applications.

In addition to the effects related to the activation pathway and conditions, it should be kept in mind that the properties of the biochar precursor will also play a key role in determining the properties (both porosity and chemical functionalities in surface) of produced activated carbons. The physicochemical properties of biochar are mainly dependent on the biomass feedstock as well as the kind and conditions of the thermo-chemical process applied.

1.5.2 *Adsorption in aqueous phase*

Several early studies were focused on using biochar-based activated carbons for the adsorption of several heavy metals in aqueous solutions, such as As(V) [Jin *et al.*, 2014], Cu(II) [Aguayo-Villarreal *et al.*, 2017; Zuo *et al.*, 2016], Hg(II) [Tan *et al.*, 2016] and Pb(II) [Ho *et al.*, 2017]. Adsorption mechanisms of metals on activated biochars mainly depend on the electrostatic interaction, ion-exchange, physisorption, and complexation and/or precipitation of functional groups [Tan *et al.*, 2017]. According to this, the most appropriate activated carbons for adsorption of heavy metals in wastewater seem to be those having a hierarchical pore size arrangement (with relatively high contributions from mesoporosity) [Aguayo-Villarreal *et al.*, 2017] with a considerable amount of oxygen-containing functionalities (e.g. hydroxyl and carboxylic groups) [Aguayo-Villarreal *et al.*, 2017; Tan *et al.*, 2017].

Biochar-based activated carbons also received special attention as promising adsorbents for removal of organic pollutants in aqueous phase. For this purpose, carbon-based adsorbents with highly developed mesoporosity are preferred, given the large size of the molecules involved (e.g. dyes and biomolecules) [Roldán *et al.*, 2016]. For the specific case of the adsorption of dyes, mesoporous carbons doped with nitrogen and/or sulfur can be synthetized through single-step HTC and salt templating. Alternatively, the introduction of heteroatoms and/or the salt templating steps can be conducted via activation of biomass-derived HTC chars. For instance, Liu *et al.* [2017] reported an excellent absorption capacity of acid orange 7 for a dual N,S-doped mesoporous carbon, which was produced via impregnation with $ZnCl_2$ of a sewage sludge-derived hydrochar.

1.5.3 *Adsorption in gaseous phase*

Adsorption of gases on biochar or biomass-derived carbons is a very promising alternative to storing them. The main advantages of these renewable carbons are their low cost, their ability to be easily regenerated and (unlike other adsorbents such as zeolites of MOFs) their hydrophobicity, which allows them to deliver an appropriate performance under wet conditions [González *et al.*, 2013].

1.5.3.1 *CO_2 capture in postcombustion*

So far, chemical absorption by aqueous-amine solutions is the most mature and industrially developed technology for CO_2 capture in power plants. However, this technology presents several drawbacks, such as corrosion rate and high energy requirements for regeneration [Creamer and Gao, 2016]. On the contrary, CO_2 capture by solid adsorbents (e.g. activated carbons) typically involves a weak bonding between CO_2 and the surface of the adsorbent, leading to lower energy requirements for the regeneration of the adsorbent. From a sustainability point of view, when the precursors used to prepare carbon-based adsorbents are lignocellulosic materials (or biochar produced from them), the attractiveness of these adsorbents is further enhanced.

For CO_2 capture in postcombustion, suitable adsorbents have to meet the following requirements [Hao *et al.*, 2013; Yang *et al.*, 2017]: (1) high CO_2 uptake at low CO_2 partial pressures (10–15 kPa), (2) high selectivity towards CO_2 over N_2, (3) fast adsorption kinetic rate, and (4) low or moderate heat of adsorption. The CO_2-over-N_2 selectivity of a porous sorbent can to a certain degree be determined by thermodynamic and kinetic reasons. Given that the molecule of CO_2 has a higher quadrupole moment $(-13.7 \times 10^{-24}$ cm^2) than that of N_2 $(-4.9 \times 10^{-26}$ cm^2), the interaction of CO_2 with the electrical field gradients of the sorbent is expected to be higher than that of N_2. From the kinetics point of view, the effective kinetic diameter within porous solids of CO_2 is a little bit smaller than that of N_2 (0.33 and 0.36 nm, respectively). Hence, the diffusion rate of CO_2 into the pores of sorbents can be higher than that of N_2, especially when the porous adsorbent has relatively uniform pores with sizes approaching the effective kinetic diameter of N_2 [Zhao *et al.*, 2015].

The main factor controlling the CO_2 uptake at low partial pressures of CO_2 (10–15 kPa) is the volume of ultra-micropores (<0.7 nm) instead of the specific surface area [Hao *et al.*, 2013]. However, developing a certain mesoporosity in the activated carbon can enhance gas transport and diffusion into the ultra-micropores by reducing the resistance to mass transfer and shortening diffusion paths [Nelson *et al.*, 2016]. In addition to the role of hierarchical porosity, modifying the surface of the adsorbent with nitrogen functional groups can result in an enhancement of the adsorption capacity of acidic gases such as CO_2 [Chen *et al.*, 2016]. Nevertheless, the effect of N-doping on the performance of CO_2 adsorption is still unclear. In fact, a marginal or event negligible effect was reported [Sevilla *et al.*, 2013]. Table 1.3 lists the CO_2 adsorption capacities at equilibrium (at 25°C and 15 kPa) recently reported in the literature for some advanced porous carbons produced from lignocellulosic precursors.

1.5.3.2 *Biogas upgrading and hydrogen purification*

In order to upgrade biogas (from anaerobic digestion) and landfill gas to natural gas quality, the main impurities (CO_2, H_2O, and H_2S) should be removed through absorption, adsorption and/or membrane separation processes. The undesired compounds can be removed sequentially starting with H_2S and H_2O and ending with the separation of CO_2 and CH_4 [Zhou *et al.*, 2017].

The adsorption of H_2S on activated carbons represents an alternative to other available technologies for the removal of H_2S from raw biogas at relatively low concentrations (e.g. below 1000 ppm_v) [Castrillon *et al.*, 2016]. Chemical activation with basic compounds (such as NaOH, KOH and K_2CO_3) can result in a significant improvement in the H_2S sorption capacity as a consequence of the involvement of a chemisorption mechanism. Impregnation of carbonaceous precursors with Fe_2O_3 was also reported to be an effective measure to increase the H_2S uptake [Bagreev *et al.*, 2001]. Recently, Hervy *et al.* [2018] reported excellent H_2S adsorption capacities (in dry conditions and room temperature) for biomass-derived chars, which were physically activated with steam. The authors also revealed that these low-cost and renewable H_2S adsorbents exhibited the following properties: high specific surface area with large mesoporous volume, alkaline pH surface, presence of mineral species on the surface (especially Ca, Al, Fe), O-containing functional groups (especially basic groups) and disorganized carbonaceous structure.

Regarding the selective adsorption of CO_2 over CH_4, pressure swing adsorption (PSA) processes have received growing attention, due to their energy and cost efficiency. Given that the PSA units usually operate at pressures ranging from 0.4 to 1.0 MPa, the performance of a given porous carbon, in terms of CO_2 uptake and selectivity towards CO_2 over CH_4, should be assessed at much higher absolute pressures than those applied in postcombustion CO_2 capture approaches. For instance, Álvarez-Gutiérrez *et al.* [2016] observed a good performance (especially at an

Table 1.3: CO_2 uptakes at 25°C and 15 kPa reported for porous carbons derived from different precursors.

References	Highest CO_2 uptake reported (mmol g^{-1})	Precursor	Preparation method
Hao *et al.* [2013]	1.10	HTC chars from grass cuttings	Physical activation with CO_2 at 800°C.
Manyà *et al.* [2018]	1.22	Pyrolyzed vine shoots at 600°C under N_2	Physical activation with CO_2 at 800°C (3 h soaking time).
Li *et al.* [2015]	1.55	Pyrolyzed rice husks at 520°C under N_2	Chemical activation with KOH: wet impregnation at a mass ratio of 1:1 and subsequent heating under N_2 at 710°C.
Yang *et al.* [2017]	1.45	Pyrolyzed coconut shells at 500°C under N_2	Chemical activation with KOH: wet impregnation at a mass ratio of 1:1 and subsequent heating under N_2 at 600°C.
Chen *et al.* [2016]	1.40	Pyrolyzed coconut shells at 500°C under N_2	Modification of the precursor by urea (a mixture at a mass ratio of 1:1 was heated in air at 320°C) and chemical activation of the resulting sample with KOH (wet impregnation at a KOH/ precursor mass ratio of 3 and subsequent heating under N_2 at 650°C).
Deng *et al.* [2014]	2.00	Pyrolyzed pine nut shells at 500°C under N_2	Chemical activation with KOH: dry mixture at a KOH/precursor mass ratio of 2 and subsequent heating under N_2 at 700°C.
González *et al.* [2013]	1.10	Almond shells	Single-step activation with CO_2 at 800°C.
Guo *et al.* [2016]	1.35	Pyrolyzed coconut shells at 500°C under N_2	Preoxidation with H_2O_2 followed by ammoxidation (by a mixture of ammonia and air at the ratio of 1:10 at 350°C for 5 h). The resulting samples were then activated with KOH (wet impregnation at a mass ratio of 1:1 and subsequent heating under N_2 at 600°C).
Parshetti *et al.* [2015]	1.20	HTC chars from empty fruit brunch at 350°C	Chemical activation with KOH: wet impregnation at a KOH/precursor mass ratio of 5 and subsequent heating under N_2 at 800°C.
Serafin *et al.* [2017]	1.25	Pomegranate peels	Single-step activation with KOH: wet impregnation at a mass ratio of 1:1 and subsequent heating under N_2 at 700°C.
Coromina *et al.* [2016]	1.50	HTC chars from Jujun grass at 350°C	Chemical activation with KOH: dry mixture at a KOH/precursor mass ratio of 2 and subsequent heating under N_2 at 700°C.

absolute pressure of 0.5 MPa) of cherry stones-derived activated carbons (physically activated with CO_2 or steam) for the separation of CO_2/CH_4 under dynamic conditions in a packed-bed.

Hydrogen can be obtained using a number of different processes, such as steam methane reforming and thermochemical conversion of coal and/ or biomass. However, the hydrogen produced also contains a high proportion of impurities, such as H_2O, CO, CO_2, CH_4 and, in some cases, N_2 [Delgado *et al.*, 2014]. The removal of these impurities can be performed using a PSA process, which is based on multilayer beds (at least 3) of different adsorbents. The first layer, which is usually composed of a hydrophilic adsorbent (e.g. alumina or silica gel), usually removes water vapor. In the second layer, the CO_2 is adsorbed using selective adsorbents, such as activated carbons; whereas the lighter compounds (e.g. CO and CH_4) are removed in a third layer, which is usually composed of zeolite-based adsorbents. So far, very few studies have focused on assessing the performance (in terms of capacity uptake and selectivity) of biochar-derived carbons for adsorption of CO_2 in a mixture of CO_2/H_2.

1.5.4 *Biochar as catalyst or catalyst support*

Carbon-based materials are widely used in the heterogeneous catalysis field because of their unique properties in terms of porosity and functional group tunability [Cao *et al.*, 2017]. As compared to non-renewable fossil fuel-derived carbonaceous materials, the production of biochar-derived materials is more sustainable and easily scalable. Thus, functionalized biochars have considerable potential to be used as direct catalysts or catalyst supports in several advanced applications. In this section, a brief overview is given on the utilization of biochar-derived materials as catalysts and/or catalyst supports in biomass upgrading processes.

1.5.4.1 *Biochar-based solid acids*

Biochar-containing sulfonic acid (SO_3H) groups, also called biochar-based solid acids, are considered as a type of metal-free catalyst that can be used in a wide variety of chemical processes. The most common way to synthesize amorphous carbon-based solid Brønsted acids is by surface sulfonation of biochar using sulfuric acid [Dehkhoda *et al.*, 2010]. Given the strong oxidation power of sulfuric acid, surface sulfonation can be accompanied by a production of oxygen-containing functionalities (e.g. –COOH and phenolic –OH), which can enhance further the catalytic activity of the biochar-based solid acids [Hara, 2010].

Biochar-based solid acid catalysts can be applied in several processes aimed to produce valuable platform chemicals, such as furfural (FF) and 5-hydroxymethylfurfural (HMF), from biomass via the hydrolysis of cellulose and hemicelluloses and further dehydration of C_6 carbohydrates [Cao *et al.*, 2017]. Several studies have already highlighted that the biochar-based solid acids represent a promising alternative to mineral acids in biomass hydrolysis and dehydration, because they are cheaper, less corrosive and easier to recycle. For instance, Hu *et al.* [2015] reported an excellent activity of a magnetic lignin-derived solid acid catalyst for the dehydration of fructose into HMF. This catalyst was prepared through a three-step impregnation–carbonization–sulfonation process using the enzymatic hydrolysis lignin residue as precursor. By impregnation of $FeCl_3$, carbonization at 400°C under N_2 atmosphere, and treatment with concentrated sulfuric acid, the resulting magnetic catalyst exhibited a relatively high surface area and contained Fe_3O_4 components, which were formed during the carbonization step. Interestingly, the Fe_3O_4 structure formed during the pyrolysis of the $FeCl_3$-preloaded biomass was not seriously affected by the sulfonation step. Thus, the lignin-based catalyst was easily separated from the reaction mixture using a magnet.

Biochar-based solid acids can also be used in biodiesel production processes. For instance, Kastner *et al.* [2012] prepared acid catalysts (by sulfonating biochar with concentrated H_2SO_4 or gaseous SO_3) with very good performance in the esterification of fatty acids. Biochar-based solid acid catalysts also showed favorable performance in the catalytic transesterification of triolein and methanol [Zeng *et al.*, 2014].

1.5.4.2 *Biochar-supported catalysts*

It is well known that the presence of inherent inorganic compounds (especially alkaline and alkaline-earth metals, such as K, Ca and Na) in the biochar matrix can catalyze several reactions involved in the thermochemical conversion of biomass (e.g. pyrolysis and gasification). Manyà *et al.* [2016] reported a significant improvement of the pyrolysis gas (i.e. higher yields of CH_4, H_2 and CO) when K_2CO_3 (5 wt.%) and CaO (5 wt.%) were added to a biomass feedstock (two-phase olive mill waste) during the slow pyrolysis of the given mixture at a peak temperature of

600°C. Wang *et al.* [2010] suggested that the *in situ* addition of K_2CO_3 promoted the decomposition of hemicelluloses, cellulose and lignin (the main biomass constituents), leading to increased yields of gaseous and biochar products at the expense of the pyrolysis liquid fraction. In other words, potassium accelerates the thermal decomposition and simultaneously promotes the secondary reactions of volatiles leading to an additional production of gas and biochar. Ren *et al.* [2014] also analyzed the *in situ* catalytic performance of a corn stover-derived biochar during the microwave pyrolysis of Douglas fir sawdust pellets at 480°C. Their results indicated that the biochar catalyst can significantly improve the quality of produced gas, since an increase in the yields of H_2, CO and CH_4 was achieved.

In addition to the *in situ* role of the inherent inorganic constituents, biochar can also be used in a downstream catalytic bed reactor to upgrade the pyrolysis vapors. For instance, Wang *et al.* [2017] used a fixed-bed of a pyrolysis char from municipal solid waste (at a temperature of 500–600°C) to reform the pyrolysis vapors coming from the pyrolysis of the same precursor. They showed a good performance of this very cheap catalyst: approximately 60% of the moisture contained in the pyrolysis vapors was consumed in the reforming process, leading to an increased gas yield with a high carbon conversion ratio into produced gas.

On the other hand, biochar-derived materials have a great potential to be used as supports to stabilize metal nanoparticles for different catalytic applications. Biochar supported metals or metal oxides can be prepared by impregnating the biochar with metal precursors. An interesting single-step route was recently suggested to synthesize biochar-supported Ni/Fe catalysts with highly dispersed metal nanoparticles by impregnation with $Fe(NO_3)_3$ and $Ni(NO_3)_2$ onto a rice husk-derived biochar and subsequent calcination at 600°C [Shen *et al.*, 2014]. This catalyst showed a good performance in the removal of tar from a biomass gasifier (tar conversion efficiencies of around 93%). The biochar support can play a dual role: (1) It can act as a reduction media, converting metal oxides into metal nanoparticles, thus enhancing the catalytic activity; (2) The biochar itself can adsorb volatile compounds contained in the pyrolysis/gasification tar, thus also enhancing tar cracking and reforming reactions [Liu *et al.*, 2015]. Yao *et al.* [2016] also produced biochar-supported Ni catalysts, by impregnation of $Ni(NO_3)_2$ onto several biochars and further calcination at 800°C, to enrich the hydrogen content of a gas produced from biomass gasification at 800°C. The best catalytic performance was observed for the Ni catalyst supported on a cotton stalk-derived biochar (H_2 content of 64 vol.% and H_2 yield of 92 mg g^{-1} of gasified biomass).

1.5.5 *Biochar-derived materials for electrochemical energy storage*

The efficient storage of energy in electrochemical energy storage devices (e.g. batteries or supercapacitors) appears to be crucial in moving from current energy supply sources (mainly based on fossil fuels) towards sustainable and renewable ones. Biochar-derived materials have a great potential to be used as electrode materials in energy storage devices, including lithium-ion batteries (LiBs), lithium-sulfur batteries (Li-SBs), sodium-ion batteries (SiBs) and supercapacitors.

1.5.5.1 *Lithium-ion batteries*

Graphite is currently the most widely used commercial anode material in lithium-ion batteries (LiBs). Nevertheless, graphite (having a theoretical specific capacity of 372 mA h g^{-1}) cannot meet the increasing demand for high capacity batteries. Despite the fact that biomass-derived carbonaceous materials have a great potential to be used as anodes in LiBs, their electrochemical performance depends, to a great extent, on the nature of the biomass precursor as well as the operating conditions adopted during the carbonization and activation processes. This fact can explain the wide variability in the performance indicators (such as specific capacity, capacity retention after a number of cycles and Coulombic efficiency) reported in the literature for several biomass-derived carbons. Some encouraging results from previous studies on developing biomass-based materials for LiBs' purposes are given in Table 1.4.

1.5.5.2 *Lithium-sulfur batteries*

With a theoretical energy density of 2600 W h kg^{-1} (3–5 times greater than that of the lithium-ion batteries [Marmorstein *et al.*, 2000]), lithium-sulfur batteries (Li-SBs) represent a very promising alternative for the next generation of rechargeable batteries. As a key component of a Li–S battery, the sulfur cathode contributes to the high energy density of the battery by delivering a remarkably high theoretical capacity of 1675 mA h g^{-1}.

Table 1.4:   Summary of the results obtained in some published studies on the performance of biomass-derived carbons as anode material for LiBs.

References	Material	Specific capacity (discharge)
Han *et al.* [2014]	Biochar from tea leaves produced by pyrolysis at 700°C, atmospheric pressure and under an N_2 atmosphere.	471 mA h g^{-1} at the 50th cycle (current density = 47.1 mA g^{-1})
Unur *et al.* [2013]	Hydrochar from hazelnut shells (produced at 250°C; soaking time = 7.5 h) thermally activated under Ar at 600°C.	300 mA h g^{-1} at the 100th cycle (current density = 300 mA g^{-1})
Yi *et al.* [2017]	Graphene-like carbon sheet (GCS) *in situ* doped GCS/Fe_3O_4 nanocomposite from soda paper-making black liquor (impregnation of black liquor with an aqueous solution of Fe^{3+}, hydrothermal treatment of the precursor solution at 180°C for 48 h, and thermal activation at 700°C under N_2).	750 mA h g^{-1} at the 1400th cycle (current density = 1000 mA g^{-1})
Wang *et al.* [2014]	Carbon-Co_3O_4 nanocomposites from pyrolyzed crawfish shells (at 700°C under N_2) and subsequent hydrothermal treatment (at 150°C for 3.5 h) of a solution of a mixture of the char precursor (previously dissolved in ethanol) and an aqueous solution of $Co(CH_3COO)_2 \cdot 4H_2O$.	1060 mA h g^{-1} at the 100th cycle (current density = 100 mA g^{-1})

However, there are still some drawbacks limiting the further development and implementation of Li-SBs (e.g. poor rate performance, capacity fading and low Coulombic efficiency) [Yang *et al.*, 2015].

Using porous carbon-based sulfur composites as cathode materials can significantly improve the performance of Li-SBs because the high electronic conductivity and electrochemical affinity for sulfur of porous carbons [Ye *et al.*, 2013]. The porosity of the carbon matrix is a key factor explaining the performance of the cathode. In this sense, it seems appropriate to develop porous carbons with hierarchical structures containing both micro- and meso-porosity. Micropores are able to adsorb the migrating polysulfides, thus avoiding their dissolution into the electrolyte. For their part, mesopores can facilitate lithium-ion transport and ensure appropriate accessibility to the electrolyte during the course of the electrochemical reactions involved [Zhang *et al.*, 2017]. Numerous studies already focused on developing high-performance cathode materials from biomass. Table 1.5 summarizes the performance of several biomass-based materials tested in Li-SB applications.

1.5.5.3 *Sodium-ion batteries*

In the last few years, sodium-ion batteries (SiBs) have gained growing attention as a low-cost alternative to LiBs because of the abundance of sodium resources throughout the world. Due to the fact that Na-ion has a larger ionic radius than that of Li-ion, one of the key challenges for SiBs' development is to use anode materials having appropriate space for intercalating and accommodating Na-ions [Gao *et al.*, 2018]. Engineered hard carbons (i.e. highly disordered and non-graphitizable carbons) are strong candidates to be used as anodes in SiBs. Using hard carbons, sodium ions can be stored in randomly stacked graphene layers and micropores.

Several hard carbons have recently been prepared from biomass sources (such as banana peels [Lotfabad *et al.*, 2014], pomelo peels [Hong *et al.*, 2014], and lignin [Li *et al.*, 2016]) and tested as anodes for SiBs. Anodes based on large surface area carbonaceous materials (with relatively large volumes of mesopores) can deliver a relatively high capacity and good rate performance, as a consequence of the broad contact between electrolytes and electrodes, which results in a fast transfer of electrons and Na-ions [Zheng *et al.*, 2016]. Nevertheless, their capacities after the first cycle can dramatically decrease, due to the formation of the solid-electrolyte interphase (SEI) [Hong *et al.*, 2014; Zheng *et al.*, 2016]. Alternatively, using hierarchical porous carbons having large micropore volumes and limited mesoporosity can lead to higher initial Coulombic efficiencies (i.e. ratio of the total charge extracted from the battery to the total charge put into the battery over a full charge cycle) with appropriate specific capacities. In other words, tuning the porous structure of the biomass-derived carbons is a key step during the preparation of anode materials for SiBs. The electrochemical performances of some biomass-derived anode materials for SiBs are summarized in Table 1.6.

Table 1.5:   Summary of the results obtained in some published studies on the performance of biomass-derived carbons as cathode materials for Li-SBs.

References	Material	Specific capacity (discharge)	Capacity retention
Guo *et al.* [2015]	Sulfur/Porous carbons nanosheet (S/PCNS) composites. PCNS were obtained from corncob by carbonization (at 500°C), wet impregnation with KOH, and thermal activation (in Ar at 900°C). Sublimed sulfur was then added to PCNS by physical mixture and subsequent heating in two steps (at 155°C for 12 h and at 300°C for 2 h).	634 mA h g^{-1} at the 50th cycle (current density = 63.4 mA g^{-1})	39.6% at the 50th cycle
Yang *et al.* [2015]	Sulfur/activated carbon (S/AC) composites. AC was obtained from apricot shells by KOH activation (at 750°C under Ar). Sublimed sulfur was then added to AC by physical mixture and subsequent heating (at 155°C for 15 and at 300°C for 0.5 h).	733 mA h g^{-1} at the 200th cycle (current density = 146.6 mA g^{-1})	61.4% at the 200th cycle
Zhang *et al.* [2014]	Sulfur/activated carbon (S/AC) composites. AC was obtained from litchi shells by carbonization (at 500°C) and subsequent KOH activation (at 900°C under Ar). Sublimed sulfur was then added to AC by physical mixture and subsequent heating at 155°C for 12 h.	665 mA h g^{-1} at the 100th cycle (current density = 800 mA g^{-1})	72.5% at the 100th cycle
Chen *et al.* [2017]	Sulfur/hierarchical porous carbon (S/LC) composites decorated by ionic surfactants. Porous carbon was obtained from lotus seedpod shells by carbonization (in Ar at 800°C) and subsequent KOH activation (at 850°C under Ar). Sulfur and ionic surfactants were then added to the porous carbon (sulfur content = 86.5%).	1160 mA h g^{-1} at the 100th cycle (current density = 116 mA g^{-1})	84.0% at the 100th cycle

Table 1.6:   Summary of the results obtained in some published studies on the performance of biomass-derived carbons as anode materials for SiBs.

References	Material	Specific capacity (discharge)	Initial Coulombic efficiency (%)
Lotfabad *et al.* [2014]	Carbonization (slow pyrolysis) of banana peels at 1100°C for 5 h under Ar. The produced carbon exhibited a very low S_{BET} (19.3 m^2 g^{-1}; deduced from the adsorption isotherm of N$_2$ at 77 K). The authors stated that this material was composed of pseudographitic arrays possessing a mean graphene interlayer spacing that was 17% dilated with respect to graphite, allowing for facile Na intercalation between the layers.	300 mA h g^{-1} at the 290th cycle (current density = 100 mA g^{-1}); with a capacity retention of 88%	67.8
Hong *et al.* [2014]	Carbonization of dried pomelo peels, previously impregnated with H$_3$PO$_4$, at 700°C for 2 h under N$_2$. A high S_{BET} of 1272 m^2 g^{-1} was reported. Pore size distribution also indicated a high proportion of mesopores in the range of 4–23 nm.	181 mA h g^{-1} at the 220th cycle (current density = 200 mA g^{-1}); with a capacity retention of 84.6%	27.0
Li *et al.* [2016]	Carbonization of a mixture of pitch and lignin (mass ration of 1:1) at 1400°C for 2 h under Ar. A very small S_{BET} of 1.34 m^2 g^{-1} was deduced from the adsorption isotherm of N$_2$ at 77 K, indicating a narrow microporous nature of the material.	226 mA h g^{-1} at the 150th cycle (current density = 30 mA g^{-1}); with a capacity retention of 89%	82.0
Wang *et al.* [2017]	Carbonization of milled mangosteen shells at 1500°C for 2 h under Ar. A very small S_{BET} of 8.9 m^2 g^{-1} was deduced from the adsorption isotherm of N$_2$ at 77 K, indicating a narrow microporous nature of the material.	330 mA h g^{-1} at the 100th cycle (current density = 20 mA g^{-1}); with a capacity retention of 98%	83.0

Table 1.7:   Summary of the results obtained in some published studies on the performance of biomass-derived carbons as electrode materials for ECs.

References	Material	Specific capacitance	Capacity retained
Peng *et al.* [2018]	A mixture of coconut shell and sewage sludge was hydrothermally treated at 220°C for 3 h. The resulted hydrochar was then activated with KOH by dry mixture at a KOH/precursor mass ratio of 3 and subsequent heating at 700°C for 3 h under N_2. A very high S_{BET} of 3000 m^2 g^{-1} was reported (total pore volume = 2.06 cm^3 g^{-1}).	420 F g^{-1} at 0.5 A g^{-1} (three-electrode configuration; 6.0 M KOH aqueous solution as electrolyte).	94.5% after 10000 cycles (current density = 5 A g^{-1})
Choi *et al.* [2018]	Dried coffee grounds were carbonized at 700°C for 2 h under Ar. The resulting char was mixed with KOH at a mass ratio KOH/char of 6 and subsequently pyrolyzed at 700°C for 2 h under Ar (S_{BET} = 2620 m^2 g^{-1}). Graphene oxide (GO) was finally dispersed at a concentration of 1.8 wt.% by chemical exfoliation of graphite.	512 F g^{-1} at 0.5 A g^{-1} (two-electrode symmetric configuration; 6.0 M KOH aqueous solution as electrolyte).	87% after 10000 cycles (current density = 5 A g^{-1})
Kang *et al.* [2018]	Rapeseed dregs and $ZnCl_2$ were mixed at a mass ratio $ZnCl_2$/precursor of 4. The resulting mixture was then heated under N_2 at 600°C for 1 h. To remove impurities, the obtained carbon was washed by 2 M HCl (S_{BET} = 1119 m^2 g^{-1}; total pore volume = 1.25 cm^3 g^{-1}).	170 F g^{-1} at 0.5 A g^{-1} (3-electrode configuration; 1.0 M H_2SO_4 aqueous solution as electrolyte).	90% after 6400 cycles (current density = 1 A g^{-1})
Gao *et al.* [2016]	Shrimp shell and KOH with a mass ratio of 1:4 were thoroughly mixed and then pyrolyzed at 700°C for 1 h in an atmosphere of Ar. To remove $CaCO_3$, the obtained carbons were dissolved in acetic acid. The $CaCO_3$-free samples were finally washed with deionized water until the solution became neutral (S_{BET} = 1113 m^2 g^{-1}; total pore volume = 0.68 cm^3 g^{-1}).	300 F g^{-1} at 0.5 A g^{-1} (3-electrode configuration; 6.0 M KOH aqueous solution as electrolyte).	93.6% after 1000 cycles (current density = 1 A g^{-1})

1.5.5.4 *Supercapacitors*

Electrochemical capacitors (ECs), also known as supercapacitors, have attracted great attention due to their high power capability, fast discharge rates, long cycling life and relatively low maintenance cost [Choi *et al.*, 2018]. So far, one of the main challenges in developing ECs is to improve the energy density without reducing the power density and/or cycling life.

Biomass-derived porous carbons are strong candidates to be used as a main electrode material in ECs. More in detail, biomass-derived carbons having large specific surface areas and well-developed hierarchical porous structure (i.e. containing micro-, meso- and macropores), and heteroatoms (doped with e.g. N, S, and B) seem to be highly appropriated for ECs [Kang *et al.*, 2018]. The presence of macropores can result in a faster ionic diffusion at the internal surface of porous carbons, whereas mesopores can act as channels to enhance the transportation of ions. For its part, a large micropore volume can provide the required locations for charge accommodation. In other words, a hierarchical pore size distribution is highly recommended in order to reach a high rate performance and power density (due to the accelerated ion transfer through the macro- and mesopores) and, at the same time, a high specific capacitance and energy density (given the large ion-accommodating surface area in the micropore region) [Feng *et al.*, 2016]. In addition, introducing heteroatoms into the carbon framework can result in a higher pseudo-capacitance caused by faradaic reactions from the heteroatom, leading to an improvement of both specific capacitance and rate performance [Gao *et al.*, 2016; Kang *et al.*, 2018]. Some examples of the performance of biomass-derived porous carbons in ECs are summarized in Table 1.7. Chapter 7 provides a more detailed overview of the potential of biomass-derived carbons as electrode components for ECs.

References

Aguayo-Villarreal, I. A., Bonilla-Petriciolet, A. and Muñiz-Valencia, R. (2017). Preparation of activated carbons from pecan nutshell and their application in the antagonistic adsorption of heavy metal ions. *J. Mol. Liq.*, 230, pp. 686–695.

Alatalo, S. M., Mäkilä, E., Repo, E., Heinonen, M., Salonen, J., Kukk, E., Sillanpää, M. and Titirici, M. M. (2016). Meso- and microporous soft templated hydrothermal carbons for dye removal from water. *Green Chem.*, 18, pp. 1137–1146.

Álvarez, M. L. (2019). Treatment of metal contaminated soils by combination of phytoremediation techniques with addition of biochar [Tratamiento de suelos contaminados por metales mediante combinación de técnicas de fitorremediación con adición de biochar (in Spanish)] (Madrid, Universidad Politécnica de Madrid).

Álvarez, M. L., Gascó, G., Plaza, C., Paz-Ferreiro, J. and Méndez, A. (2017). Hydrochars from biosolids and urban wastes as substitute materials for peat. *Land Degrad. Dev.*, 28(7), pp. 2268–2276.

Álvarez-Gutiérrez, N., García, S., Gil, M. V., Rubiera, F. and Pevida, C. (2016). Dynamic performance of biomass-based carbons for CO2/CH4 separation. Approximation to a pressure swing adsorption process for biogas upgrading. *Energy Fuels*, 30, pp. 5005–5015.

Bagreev, A., Bashkova, S., Locke, D. G. and Bandosz, T. J. (2001). Sewage sludge-derived materials as efficient adsorbents for removal of hydrogen sulfide. *Environ. Sci. Technol.*, 35, pp. 1537–1543.

Bruun, E. W., Hauggaard-Nielsen, H., Ibrahim, N., Egsgaard, H., Ambus, P., Jensen, P. A. and Dam-Johansen, K. (2011). Influence of fast pyrolysis temperature on biochar labile fraction and short-term carbon loss in a loamy soil. *Biomass Bioenergy*, 35, pp. 1182–1189.

Cao, X., Sun, S. and Sun, R. (2017). Application of biochar-based catalysts in biomass upgrading: A review. *RSC Adv.*, 7, pp. 48793–48805.

Cárdenas-Aguiar, E., Gascó, G., Paz-Ferreiro, J. and Méndez, A. (2019). Thermogravimetric analysis and carbon stability of chars produced from slow pyrolysis and hydrothermal carbonization of manure waste. *J. Anal. Appl. Pyrolysis*, 140, pp. 434–443.

Castrillon, M. C., Moura, K. O., Alves, C. A., Bastos-Neto, M., Azevedo, D. C. S., Hofmann, J., Möllmer, J., Einicke, W. D. and Gläser, R. (2016). CO_2 and H_2S Removal from CH_4-Rich Streams by Adsorption on Activated Carbons Modified with K_2CO_3, NaOH, or Fe_2O_3. *Energy Fuels*, 30, pp. 9596–9604.

Chen, J., Yang, J., Hu, G., Hu, X., Li, Z., Shen, S., Radosz, M. and Fan, M. (2016). Enhanced CO_2 Capture Capacity of Nitrogen-Doped Biomass-Derived Porous Carbons. *ACS Sustain. Chem. Eng.*, 4, pp. 1439–1445.

Chen, M., Jiang, S., Cai, S., Wang, X., Xiang, K., Ma, Z., Song, P. and Fisher, A. C. (2017). Hierarchical porous carbon modified with ionic surfactants as efficient sulfur hosts for the high-performance lithium-sulfur batteries. *Chem. Eng. J.*, 313, pp. 404–414.

Choi, J. H., Lee, C., Cho, S., Moon, G. D., Kim, B. Su, Chang, H. and Jang, H. D. (2018). High capacitance and energy density supercapacitor based on biomass-derived activated carbons with reduced graphene oxide binder. *Carbon*, 132, pp. 16–24.

Coromina, H. M., Walsh, D. A. and Mokaya, R. (2016). Biomass-derived activated carbon with simultaneously enhanced CO_2 uptake for both pre and post combustion capture applications. *J. Mater. Chem. A*, 4, pp. 280–289.

Creamer, A. E. and Gao, B. (2016). Carbon-Based adsorbents for post-combustion CO_2 capture: A critical review. *Environ. Sci. Technol.*, 50, pp. 7276–7289.

Cuong, T. N. C., Farrella, C., Kristiansen, P. E. and Rayner, J. P. (2014). Biochar makes green roof substrates lighter and improves water supply to plants. *Ecol. Eng.*, 71, pp. 368–374.

Dehkhoda, A. M., West, A. H. and Ellis, N. (2010). Biochar based solid acid catalyst for biodiesel production. *Appl. Catal. A Gen.*, 382, pp. 197–204.

Delgado, J. A., Águeda, V. I., Uguina, M. A., Sotelo, J. L., Brea, P. and Grande, C. A. (2014). Adsorption and diffusion of H_2, CO, CH_4, and CO_2 in BPL activated carbon and 13X zeolite: Evaluation of performance in pressure swing adsorption hydrogen purification by simulation. *Ind. Eng. Chem. Res.*, 53, pp. 15414–15426.

Delinat Institute (2013). European Biochar Certificate. Guidelines for biochar production (Switzerland).

Deng, S., Wei, H., Chen, T., Wang, B., Huang, J. and Yu, G. (2014). Superior CO_2 adsorption on pine nut shell-derived activated carbons and the effective micropores at different temperatures. *Chem. Eng. J.*, 253, pp. 46–54.

Feng, H., Hu, H., Dong, H., Xiao, Y., Cai, Y., Lei, B., Liu, Y. and Zheng, M. (2016). Hierarchical structured carbon derived from bagasse wastes: A simple and efficient synthesis route and its improved electrochemical properties for high-performance supercapacitors. *J. Power Sources*, 302, pp. 164–173.

Gao, F., Qu, J., Geng, C., Shao, G. and Wu, M. (2016). Self-templating synthesis of nitrogen-decorated hierarchical porous carbon from shrimp shell for supercapacitors. *J. Mater. Chem. A*, 4, pp. 7445–7452.

Gao, M., Pan, S. Y., Chen, W. C. and Chiang, P. C. (2018). A cross-disciplinary overview of naturally derived materials for electrochemical energy storage. *Mater. Today Energy*, 7, pp. 58–79.

Gascó, G., Álvarez, M. L., Paz-Ferreiro, J., San Miguel, G. and Méndez, A. (2018). Valorization of biochars from pinewood gasification and municipal solid waste torrefaction as peat substitutes. *Environ. Sci. Pollut. Res.*, 25(26), pp. 26461–26469.

Gascó, G., Paz-Ferreiro, J., Álvarez, M. L., Saa, A. and Méndez, A. (2018). Biochars and hydrochars prepared by pyrolysis and hydrothermal carbonisation of pig manure. *Waste Manage.*, 79, pp. 395–403.

Gascó, G., Paz-Ferreiro, J., Cely, P., Plaza, P. and Méndez, A. (2016). Influence of pig manure and its biochar on soil CO_2 emissions and soil enzymes. *Ecol. Eng.*, 95, pp. 19–24.

Guerrero, F. (1987). Study of the physical and chemical properties of Spanish peatlands and their possible agricultural use [Estudio de las propiedades físicas y químicas de algunas turbas españolas y su posible aprovechamiento agrícola (in Spanish)]. (Madrid, Universidad Politécnica de Madrid).

González, A. S., Plaza, M. G., Rubiera, F. and Pevida, C. (2013). Sustainable biomass-based carbon adsorbents for post-combustion CO_2 capture. *Chem. Eng. J.*, 230, pp. 456–465.

Guo, J., Zhang, J., Jiang, F., Zhao, S., Su, Q. and Du, G. (2015). Microporous carbon nanosheets derived from corncobs for lithium-sulfur batteries. *Electrochim. Acta.*, 176, pp. 853–860.

Guo, L., Yang, J., Hu, G., Hu, X., Wang, L., Dong, Y., DaCosta, H. and Fan, M. (2016). Role of hydrogen peroxide preoxidizing on CO_2 adsorption of nitrogen-doped carbons produced from coconut shell. *ACS Sustain. Chem. Eng.*, 4, pp. 2806–2813.

Han, S. W., Jung, D. W., Jeong, J. H. and Oh, E. S. (2014). Effect of pyrolysis temperature on carbon obtained from green tea biomass for superior lithium ion battery anodes. *Chem. Eng., J.*, 254, pp. 597–604.

Hao, W., Björkman, E., Lilliestråle, M. and Hedin, N. (2013). Activated carbons prepared from hydrothermally carbonized waste biomass used as adsorbents for CO_2. *Appl. Energy*, 112, pp. 526–532.

Hara, M. (2010). Biomass conversion by a solid acid catalyst. *Energy Environ. Sci.* 3, pp. 601–607.

He, C., Giannis, A. and Wang, J. Y. (2013). Conversion of sewage sludge to clean solid fuel using hydrothermal carbonization: Hydrochar fuel characteristics and combustion behaviour. *Appl. Energy*, 111, pp. 257–266.

Hervy, M., Pham Minh, D., Gérente, C., Weiss-Hortala, E., Nzihou, A., Villot, A. and Le Coq, L. (2018). H_2S removal from syngas using wastes pyrolysis chars. *Chem. Eng. J.*, 334, pp. 2179–2189.

Hiltzl, M., Corma, A., Pomares, F. and Renz, M. (2015). The hydrothermal carbonization (HTC) plant as decentral biorefinery for wet biomass. *Catal. Today*, 257, pp. 154–159.

Ho, S.-H., Chen, Y., Yang, Z., Nagarajan, D., Chang, J.-S. and Ren, N. (2017). High-efficiency removal of lead from wastewater by biochar derived from anaerobic digestion sludge. *Bioresour. Technol.*, 246, pp. 142–149.

Hong, K., Qie, L., Zeng, R., Yi, Z., Zhang, W., Wang, D., Yin, W., Wu, C., Fan, Q., Zhang, W.-X. and Huang, Y. (2014). Biomass derived hard carbon used as a high performance anode material for sodium ion batteries. *J. Mater. Chem. A*, 2, p. 12733.

Hu, L., Tang, X., Wu, Z., Lin, L., Xu, J., Xu, N. and Dai, B. (2015). Magnetic lignin-derived carbonaceous catalyst for the dehydration of fructose into 5-hydroxymethylfurfural in dimethylsulfoxide. *Chem. Eng. J.*, 263, pp. 299–308.

International Biochar Initiative. (2015). Standardized product definition and product testing guidelines for biochar that is used in soil (International biochar initiative).

Jin, H., Capareda, S., Chang, Z., Gao, J., Xu, Y. and Zhang, J. (2014). Biochar pyrolytically produced from municipal solid wastes for aqueous As(V) removal: Adsorption property and its improvement with KOH activation. *Bioresour. Technol.*, 169, pp. 622–629.

Kang, X., Zhu, H., Wang, C., Sun, K. and Yin, J. (2018). Biomass derived hierarchically porous and heteroatom-doped carbons for supercapacitors. *J. Colloid Interface Sci.*, 509, pp. 369–383.

Kastner, J. R., Miller, J., Geller, D. P., Locklin, J., Keith, L. H. and Johnson, T. (2012). Catalytic esterification of fatty acids using solid acid catalysts generated from biochar and activated carbon. *Catal. Today*, 190, pp. 122–132.

Lehmann, J. and Joseph, S. (2009). Biochar for environmental management: An introduction, In: Lehmann, J. and Joseph, J. (eds.), Biochar for Environmental Management (Earthscan, London), pp. 1–9.

Li, D., Ma, T., Zhang, R., Tian, Y. and Qiao, Y. (2015). Preparation of porous carbons with high low-pressure CO_2 uptake by KOH activation of rice husk char. *Fuel*, 139, pp. 68–70.

Li, M., Liu, C., Cao, H., Zhao, H., Zhang, Y. and Fan, Z. (2014). KOH self-templating synthesis of three-dimensional hierarchical porous carbon materials for high performance supercapacitors. *J. Mater. Chem. A*, 2, pp. 14844–14851.

Li, Y., Hu, Y.-S., Li, H., Chen, L. and Huang, X. (2016). A superior low-cost amorphous carbon anode made from pitch and lignin for sodium-ion batteries. *J. Mater. Chem. A*, 4, pp. 96–104.

Liang, C., Zhu, X., Fu, S., Méndez, A., Gascó, G. and Paz-Ferreiro, J. (2014). Biochar alters the resistance and resilience to drought in a tropical soil. *Environ. Res. Lett.*, 9(6), 064013.

Liu, T., Li, Y., Peng, N., Lang, Q., Xia, Y., Gai, C., Zheng, Q. and Liu, Z. (2017). Heteroatoms doped porous carbon derived from hydrothermally treated sewage sludge: Structural characterization and environmental application. *J. Environ. Manage.*, 197, pp. 151–158.

Liu, W.-J., Jiang, H. and Yu, H.-Q. (2015). Development of biochar-based functional materials: Toward a sustainable platform carbon material. *Chem. Rev.*, 115, pp. 12251–12285.

Lotfabad, E. M., Ding, J., Cui, K., Kohandehghan, A., Kalisvaart, W. P., Hazelton, M. and Mitlin, D. (2014). High-density sodium and lithium ion battery anodes from banana peels. *ACS Nano*, 8, pp. 7115–7129.

Lu, H., Li, Z., Fu, S., Méndez, A., Gascó, G. and Paz-Ferreiro, J. (2015). Combining phytoextraction and biochar addition improves soil biochemical properties in a soil contaminated with Cd. *Chemosphere*, 119, pp. 209–216.

Manyà, J. J., Alvira, D., Azuara, M., Bernin, D. and Hedin, N. (2016). Effects of pressure and the addition of a rejected material from municipal waste composting on the pyrolysis of two-phase olive mill waste. *Energy Fuels*, 30, pp. 8055–8064.

Manyà, J. J., González, B., Azuara, M. and Arner, G. (2018). Ultra-microporous adsorbents prepared from vine shoots-derived biochar with high CO_2 uptake and CO_2/N_2 selectivity. *Chem. Eng. J.*, 345, pp. 631–639.

Marmorstein, D., Yu, T. H., Striebel, K. A., McLarnon, F. R., Hou, J. and Cairns, E. J. (2000). Electrochemical performance of lithium/sulfur cells with three different polymer electrolytes. *J. Power Sources*, 89, pp. 219–226.

Méndez, A., Barriga, S., Fidalgo, J. M. and Gascó, G. (2009). Adsorbent materials from paper industry waste materials and their use in Cu(II) removal from water. *J. Hazard. Mater.*, 165, pp. 736–743.

Méndez, A., Gómez, A., Paz-Ferreiro, J. and Gascó, G. (2012). Effects of sewage sludge biochar on plant metal availability after application to a Mediterranean soil. *Chemosphere*, 89, pp. 1354–1359.

Méndez, A., Paz-Ferreiro, J., Gil, E. and Gascó, G. (2015). The effect of paper sludge and biochar addition on brown peat and coir based growing media properties. *Sci. Hortic.*, 193, pp. 225–230.

Méndez, A., Tarquis, A. M., Saa-Requejo, A., Guerrero, F. and Gascó, G. (2013). Influence of pyrolysis temperature on composted sewage sludge biochar priming effect in a loamy soil. *Chemosphere*, 93(4), pp. 668–676.

Mullen, C. A., Boateng, A. A., Goldberg, N. M., Lima, I. M., Laird, D. A. and Hicks, K. B. (2010). Bio-oil and bio-char production from corn cobs and stover by fast pyrolysis. *Biomass Bioenergy*, 34, pp. 67–74.

Nelson, K. M., Mahurin, S. M., Mayes, R. T., Williamson, B., Teague, C. M., Binder, A. J., Baggetto, L., Veith, G. M. and Dai, S. (2016). Preparation and CO_2 adsorption properties of soft-templated mesoporous carbons derived from chestnut tannin precursors. *Microporous Mesoporous Mater.*, 222, pp. 94–103.

Nieto, A., Gascó, G., Paz-Ferreiro, J., Fernández, J. M., Plaza, C. and Méndez, A. (2016). The effect of pruning waste and biochar addition on brown peat based growing media properties. *Sci. Hortic.*, 199, pp. 142–148.

Parshetti, G. K., Chowdhury, S. and Balasubramanian, R. (2015). Biomass derived low-cost microporous adsorbents for efficient CO_2 capture. *Fuel*, 148, pp. 246–254.

Paz-Ferreiro, J., Gascó, G., Gutiérrez, B. and Méndez, A. (2012). Soil biochemical activities and the geometric mean of enzyme activities after application of sewage sludge and sewage sludge biochar to soil. *Biol. Fert. Soils*, 48, pp. 511–517.

Paz-Ferreiro, J., Lu, H., Fu, S., Méndez, A. and Gascó, G. (2014). Use of phytoremediation and biochar to remediate heavy metal polluted soils: A review. *Solid Earth*, 5, pp. 65–75.

Paz-Ferreiro, J., Méndez, A. and Gascó, G. (2016). Application of biochar for soil biological improvement, In: Guo, M., He, Z. and Uchimiya, S. M. (eds.), Agricultural and environmental applications of biochar: Advances and barriers (SSSA, Madison), pp. 145–174.

Peng, L., Liang, Y., Dong, H., Hu, H., Zhao, X., Cai, Y., Xiao, Y., Liu, Y. and Zheng, M. (2018). Super-hierarchical porous carbons derived from mixed biomass wastes by a stepwise removal strategy for high-performance supercapacitors. *J. Power Sources*, 377, pp. 151–160.

Ren, S., Lei, H., Wang, L., Bu, Q., Chen, S. and Wu, J. (2014). Hydrocarbon and hydrogen-rich syngas production by biomass catalytic pyrolysis and bio-oil upgrading over biochar catalysts. *RSC Adv.*, 4, pp. 10731–10737.

Roldán, L., Marco, Y. and García-Bordejé, E. (2016). Bio-sourced mesoporous carbon doped with heteroatoms (N,S) synthesised using one-step hydrothermal process for water remediation. *Microporous Mesoporous Mater.*, 222, pp. 55–62.

Serafin, J., Narkiewicz, U., Morawski, A. W., Wróbel, R. J. and Michalkiewicz, B. (2017). Highly microporous activated carbons from biomass for CO_2 capture and effective micropores at different conditions. *J. CO_2 Util.*, 18, pp. 73–79.

Sevilla, M., Parra, J. B. and Fuertes, A. B. (2013). Assessment of the role of micropore size and N-doping in CO_2 capture by porous carbons. *ACS Appl. Mater. Interfaces*, 5, pp. 6360–6368.

Shen, Y., Zhao, P., Shao, Q., Ma, D., Takahashi, F. and Yoshikawa, K. (2014). In-situ catalytic conversion of tar using rice husk char-supported nickel-iron catalysts for biomass pyrolysis/gasification. *Appl. Catal. B Environ.*, 152–153, pp. 140–151.

Sopeña, F., Semple, K., Sohi, S. and Bending, G. (2012). Assessing the chemical and biological accessibility of the herbicide isoproturon in soil amended with biochar. *Chemosphere*, 88, pp. 77–83.

Steiner, C. and Harttung, T. (2014). Biochar as a growing media additive and peat substitute. *Solid Earth*, 5, pp. 995–999.

Strack, M. (2008). *Peatlands and Climate Change* (Jyväskylä, International Peat Society).

Tan, G., Sun, W., Xu, Y., Wang, H. and Xu, N. (2016). Sorption of mercury (II) and atrazine by biochar, modified biochars and biochar based activated carbon in aqueous solution. *Bioresour. Technol.*, 211, pp. 727–735.

Tan, X., Liu, S.-B., Liu, Y., Gu, Y., Zeng, G., Hu, X., Wang, X., Liu, S.-H. and Jiang, L. (2017). Biochar as potential sustainable precursors for activated carbon production: Multiple applications in environmental protection and energy storage. *Bioresour. Technol.*, 227, pp. 359–372.

Tian, Y, Sun, X., Li, S., Wang, H., Wang, L., Cao, J. and Zhang, L. (2012). Biochar made from green waste as peat substitute in growth media for Calathea rotundifola cv. Fasciata. *Sci. Hortic.*, 143, pp. 15–18.

Titirici, M.-M., White, R. J., Falco, C. and Sevilla, M. (2012). Black perspectives for a green future: Hydrothermal carbons for environment protection and energy storage. *Energy Environ. Sci.*, 5, pp. 6796–6822.

Uchimiya, M., Chang, S. C. and Klasson, K. T. (2011). Screening biochars for heavy metal retention in soil: Role of oxygen functional groups. *J. Hazard. Mater.*, 190, pp. 432–444.

Unur, E., Brutti, S., Panero, S. and Scrosati, B. (2013). Nanoporous carbons from hydrothermally treated biomass as anode materials for lithium ion batteries. *Microporous Mesoporous Mater.*, 174, pp. 25–33.

Wang, K., Jin, Y., Sun, S., Huang, Y., Peng, J., Luo, J., Zhang, Q., Qiu, Y., Fang, C. and Han, J. (2017). Low-cost and high-performance hard carbon anode materials for sodium-ion batteries. *ACS Omega*, 2, pp. 1687–1695.

Wang, L., Zheng, Y., Wang, X., Chen, S., Xu, F., Zuo, L., Wu, J., Sun, L., Li, Z., Hou, H. and Song, Y. (2014). Nitrogen-doped porous carbon/Co_3O_4 nanocomposites as anode materials for lithium-ion batteries. *ACS Appl. Mater. Interfaces*, 6, pp. 7117–7125.

Wang, N., Chen, D., Arena, U. and He, P. (2017). Hot char-catalytic reforming of volatiles from MSW pyrolysis. *Appl. Energy*, 191, pp. 111–124.

Wang, Z., Wang, F., Cao, J. and Wang, J. (2010). Pyrolysis of pine wood in a slowly heating fixed-bed reactor: Potassium carbonate versus calcium hydroxide as a catalyst. *Fuel Process. Technol.*, 91, pp. 942–950.

Wang, T., Camps-Arbestain, M., Hedley, M. and Bishop, P. (2012). Chemical and bioassay characterisation of nitrogen availability in biochar produced from dairy manure and biosolids. *Org. Geochem.*, 51, pp. 45–54.

Wei, H., Deng, S., Hu, B., Chen, Z., Wang, B., Huang, J. and Yu, G. (2012). Granular bamboo-derived activated carbon for high CO_2 adsorption: The dominant role of narrow micropores. *ChemSusChem*, 5, pp. 2354–2360.

Wolf, D., Amonette, J. E., Street-Perrot, F. A. and Lehmann, J. (2010). Sustainable biochar to mitigate global climate change. *Nat. Commun.*, 1, pp. 1–9.

Yang, J., Yue, L., Hu, X., Wang, L., Zhao, Y., Lin, Y., Sun, Y., DaCosta, H. and Guo, L. (2017). Efficient CO_2 capture by porous carbons derived from coconut shell. *Energy Fuels*, 31, pp. 4287–4293.

Yang, K., Gao, Q., Tan, Y., Tian, W., Zhu, L. and Yang, C. (2015). Microporous carbon derived from Apricot shell as cathode material for lithium-sulfur battery. *Microporous Mesoporous Mater.*, 204, pp. 235–241.

Yao, D., Hu, Q., Wang, D., Yang, H., Wu, C., Wang, X. and Chen, H. (2016). Hydrogen production from biomass gasification using biochar as a catalyst/ support. *Bioresour. Technol.*, 216, pp. 159–164.

Ye, H., Yin, Y.-X., Xin, S. and Guo, Y.-G. (2013). Tuning the porous structure of carbon hosts for loading sulfur toward long lifespan cathode materials for Li–S batteries. *J. Mater. Chem., A* 1, p. 6602.

Yi, X., He, W., Zhang, X., Yue, Y., Yang, G., Wang, Z., Zhou, M. and Wang, L. (2017). Graphene-like carbon sheet/Fe_3O_4 nanocomposites derived from soda papermaking black liquor for high performance lithium ion batteries. *Electrochim. Acta*, 232, pp. 550–560.

Zeng, D., Liu, S., Gong, W., Wang, G., Qiu, J. and Chen, H. (2014). Synthesis, characterization and acid catalysis of solid acid from peanut shell. *Appl. Catal. A Gen.*, 469, pp. 284–289.

Zhang, S., Zheng, M., Lin, Z., Li, N., Liu, Y., Zhao, B., Pang, H., Cao, J., He, P. and Shi, Y. (2014). Activated carbon with ultrahigh specific surface area synthesized from natural plant material for lithium–sulfur batteries. *J. Mater. Chem., A* 2, pp. 15889–15896.

Zhang, X., Xie, H., Kim, C. S., Zaghib, K., Mauger, A. and Julien, C. M. (2017). Advances in lithium-sulfur batteries. *Mater. Sci. Eng. R Reports*, 121, pp. 1–29.

Zhao, Y., Liu, X. and Han, Y. (2015). Microporous carbonaceous adsorbents for CO_2 separation via selective adsorption. *RSC Adv.*, 5, pp. 30310–30330.

Zheng, P., Liu, T. and Guo, S. (2016). Micro-nano structure hard carbon as a high performance anode material for sodium-ion batteries. *Sci. Rep.*, 6, p. 35620.

Zhou, K., Chaemchuen, S. and Verpoort, F. (2017). Alternative materials in technologies for biogas upgrading via CO_2 capture. *Renew. Sustain. Energy Rev.*, 79, pp. 1414–1441.

Zuo, X., Liu, Z. and Chen, M. (2016). Effect of H_2O_2 concentrations on copper removal using the modified hydrothermal biochar. *Bioresour. Technol.*, 207, pp. 262–267.

Chapter 2

Biochar Production via Pyrolysis

Frederik Ronsse[*,§], Ondřej Mašek[†] and Joan J. Manyà[‡]

*Department of Green Chemistry and Technology,
Ghent University, Ghent, Belgium*

†*UK Biochar Research Centre, The University of Edinburgh,
Edinburgh, UK*

‡*Escuela Politécnica Superior, Aragón Institute
of Engineering Research (i3A),
Universidad de Zaragoza, Huesca, Spain*

§*frederik.ronsse@ugent.be*

Abstract

The chapter starts with an overview of the different thermochemical conversion techniques for dry biomass, including pyrolysis and gasification. As biochars' physicochemical characteristics can vary significantly depending on the type of production technology being used, the production parameters applied and the available biomass feedstock, the second part of this chapter delves deeper into these relationships. The known relationships between feedstock, conversion conditions and biochar properties form the foundation for the production of tailor-made biochars.

2.1 Introduction

To obtain biochar from dry biomass, the latter needs to undergo a thermo-chemical conversion process [Lehmann and Joseph, 2009; Masek *et al.*, 2013; Qian *et al.*, 2015]. Although a large variety of distinct thermo-chemical conversion processes exist, they all have in common that the biomass feedstock is subjected to high temperatures (above 300°C). This chapter will only consider those conversion processes which are carried on in an atmosphere that contains little to no oxygen — combustion, which is a thermochemical conversion with an excess of oxygen, is not further elaborated upon as it does not produce char. Distinction between different thermochemical conversion processes is made on the basis of process conditions (temperature, heating rate, reaction time, etc.) and the resulting differences in product yields. An overview of the major thermo-chemical conversion processes for dry biomass which, to a varying extent, all result in the (co)production of char are listed in Table 2.1. Throughout this chapter the terms biochar and char are used interchangeably, as they refer to the same material — the only difference being the spe-cific application for which biochar is destined. Distinction can be made between pyrolysis (either fast, slow or intermediate), gasification and torrefaction.

In pyrolysis, the thermal decomposition of biomass is performed in an oxygen-free or oxygen-limited atmosphere. Hereby the biomass feedstock is broken down into three product fractions: (a) a mixture of non-condens-able gases, (b) vapors which after condensation give rise to pyrolysis liquids (also known as pyrolysis oil or bio-oil) and (c) a solid residue, better known as char [Mohan *et al.*, 2006]. The yield of each of these three pyrolysis products highly depends on the type of biomass feedstock used, as well as the prevailing pyrolysis process conditions [Bridgwater and Peacocke, 2000; Lu *et al.*, 2009].

The most important process variable by which pyrolysis processes are differentiated is the heating rate. In fast pyrolysis, biomass particles are subjected to high heating rates, typically in the order of several hundreds of degree centigrade per minute although there is no universal criterion by which required heating rates are defined in fast pyrolysis. In order to achieve these high heating rates and considering the rather limited ther-mal conductivity in biomass, the feedstock particles have to be kept small enough, typically smaller than 3 mm. In fast pyrolysis, these high heating rates are combined with short vapor residence times (i.e. the time

Table 2.1: Typical process conditions and product yields in wt.%, obtained from different types of dry thermochemical conversion processes. Note that product yield data are typical results one can obtain when using low-ash lignocellulosic feedstock (like wood).

	Torrefaction	Slow pyrolysis	Intermediate pyrolysis	Fast pyrolysis	Gasification
Temperature	<300°C	>400°C	420–600°C	~500°C	600–1800°C
Heating rate	—	<80°C min^{-1}	Several 10's °C min^{-1}	Fast, up to 1000°C min^{-1}	—
Reaction time	<2h	Hours to days	Minutes	Few seconds	—
Pressure	Atmospheric	Atmospheric (or elevated up to 1 MPa)	Atmospheric	Atmospheric (and vacuum)	Atmospheric, pressurized up to 8 MPa
Medium	Oxygen-free	Oxygen-free or oxygen-limited	Oxygen-free	Oxygen-free	Oxygen-limited (air or steam/oxygen)
Bio-oil yield (wt.%)	5	30	50	75	5
Permanent gases yield (wt.%)	15	35	20	13	85
Char (solids) yield (wt.%)	80	35	30	12	10

Sources: Data from Bridgwater [2012]; Van der Stelt *et al.* [2011]; Williams and Besler [1996]; Bain and Broer [2011]; Nachenius *et al.* [2013]; Yang *et al.* [2014].

the produced vapors spend in the reactor before being condensed into liquids), up to several seconds, in order to suppress secondary vapor cracking reactions [Czernik and Bridgwater, 2004; Balat *et al.*, 2009; Bridgwater, 2012].

Typically, when a low-ash lignocellulosic biomass feedstock is used — for instance, pine wood — then bio-oil yields usually range between 60 and 70 wt.% (on dry feedstock basis), while char yields are between 12 and 15 wt.% — the remainder of the mass balance is made up of the non-condensable gases [Isahak *et al.*, 2012].

In contrast, slow pyrolysis processes are characterized by heating rates much lower than those employed in fast pyrolysis. As a consequence, feedstock particle size is no longer critical from the perspective of heat transfer rates, and typically larger feedstock particle sizes are used in slow pyrolysis, up to several cm, when compared to fast pyrolysis. Additionally, in slow pyrolysis the contact between the pyrolysis vapors and the reacting biomass as well as char can be promoted in order to ensure a high char yield — more details on char forming chemistry are provided in Section 2.2. In literature, often the term "carbonization" is synonymous to slow pyrolysis, usually in the context of producing charcoal (i.e. the char product is destined to be used as a fuel). A special type of slow pyrolysis is torrefaction (see Table 2.1), as it shares similar process conditions as slow pyrolysis for the production of biochar or charcoal, the only difference being the maximum temperature at which torrefaction is carried out. Whereas slow pyrolysis is characterized by temperatures above 400°C to ensure complete conversion of the biomass feedstock, torrefaction employs temperatures typically below 300°C [Basu, 2013]. These milder temperatures only induce a partial decomposition of the biomass, with limited devolatilization. The result of biomass torrefaction is a solid product with increased energy density, higher hydrophobicity, improved grindability and lower susceptibility to biodegradation compared to the parent biomass feedstock [Nachenius *et al.*, 2013]. As such, torrefaction is a biomass pretreatment that aims to improve the physicochemical properties of the biomass for solid fuel purposes.

Intermediate pyrolysis is a more recent technique and has emerged to provide an improved co-production of both bio-oil and biochar. Fast pyrolysis solely focuses on the production of bio-oil to the maximum extent possible and, because of the endothermicity (see below), the process typically requires the char to be burned to provide process heat. Meanwhile, slow pyrolysis is tailored towards maximum char production, while the obtained pyrolysis liquids are often lacking in quality in respect to fuel applications because of their high water content. As such, employing process conditions (in terms of temperature, heating rate and reaction time) somewhat between fast and slow pyrolysis, significant yields of pyrolysis liquids can still be obtained of which the organic fraction can be used for fuel applications, while producing char in yields slightly below those expected in slow pyrolysis conditions [Yang *et al.*, 2014].

Gasification differs from all pyrolysis processes in its use of higher temperatures (>700°C) and the use of an oxidizer, being pure O_2, air, CO_2, steam or mixtures thereof. The oxidizer is supplied in sub-stoichiometric quantities, the equivalence ratio typically being between 0.25 and 0.33. Gasification is different from pyrolysis (fast, slow or intermediate) as it primarily targets the production of a combustible gas, better known as producer gas, which is a mixture of non-condensable gases such as H_2, CO, CH_4, CO_2 and some light hydrocarbons. When producer gas is purified to only contain H_2 and CO, it is referred to as syngas or synthesis gas. A wide variety of applications of producer gas and syngas have been developed and are in commercial operation, largely because syngas (or producer gas) can also be obtained from fossil resources like coal and natural gas. Current applications include the use of syngas (or producer gas) to power internal combustion engines or gas turbines, as boiler fuel or as a feedstock in the production of liquid fuels and chemical intermediates [McKendry, 2002]. Nonetheless, some gasification techniques (specifically those that do not completely combust the char) still can produce certain quantities of char, with low yields typically less than 10 wt.%.

All thermochemical conversion processes summarized in Table 2.1 require relatively dry feedstock, with a moisture content below 30 wt.%, but moisture contents in the vicinity of 10 wt.% are preferred. Feedstock moisture is detrimental to the energy balance of the conversion process: the more moisture present, the more heat is required to evaporate the moisture during the heating of the biomass towards its reaction temperature [Antal and Grønli, 2003]. Additionally, the resulting (feedstock) moisture in the pyrolysis vapors will reduce the overall calorific content of the vapors, which is undesired when the vapors are burned to provide process energy, as well as when the vapors are condensed into pyrolysis oil. If biomass feedstocks with a high moisture content have to be converted, drying is required prior to conversion, which is costly in terms of energy. Alternatively, wet biomass feedstocks may be converted in an aqueous environment at elevated temperatures and elevated pressures (i.e. the saturation vapor pressure corresponding to the temperature operated at). These processes, collectively called hydrothermal processes, do not require a drying step prior to conversion and water plays an active role, not only as a solvent, but also as a reactant [Kruse and Dahmen, 2015]. Hydrothermal conversion is however outside of the scope of this chapter.

2.2 Pyrolysis Chemistry and Char Formation

Plant-based biomass mainly consists, apart from water, of structural cell wall components, which include macrobiopolymers like cellulose, hemicellulose and lignin. Additional non-structural components are also present, which include minerals, sugars, lipids, waxes and proteins, collectively known as extractives, as they can be isolated out of the biomass by means of solvent extraction, unlike the structural macrobiopolymers [Mohan *et al.*, 2006]. When biomass is heated, each of the biomass constituents will undergo a combination of chemical transformation and/or partial or complete devolatilization. Upon heating, the majority of non-structural components are the first to be removed and/or transformed: water, both in its free and bound state, is the first component to be completely removed when heating biomass to temperatures above 160°C [Grønli and Melaaen, 2000; Zhou *et al.*, 2013]. Next, at higher temperatures, the majority of extractives (excluding the minerals) are devolatilized or decomposed [Yang *et al.*, 2014]. In terms of structural biomass constituents, the first macrobiopolymer to decompose is the hemicellulose. The latter is a polysaccharide, mainly built out of C5 and C6 sugars [Saha, 2003]. The degree of polymerization is low, only up to several 100s of monosaccharide units, and the polymer is branched. As a result, hemicellulose is amorphous and unlike cellulose not stabilized by hydrogen bonds. The consequence thereof is that hemicellulose is rather unstable at higher temperatures, and decomposes at temperatures between 220°C and 315°C [Stefanidis *et al.*, 2014; Yang *et al.*, 2007, 2014; Zhou *et al.*, 2013]. As mentioned before, this temperature range is the one also employed in torrefaction processes, and hence, the only macrobiopolymer to be significantly degraded is the hemicellulose. Cellulose, on the other hand, is a polysaccharide that uniquely consists of glucose units, in an unbranched polymer with degrees of polymerization up to 10,000 [White *et al.*, 2011]. These linear cellulose polymers can align themselves to one another, while being stabilized by Van der Waals forces and hydrogen bonds. As such, cellulose is predominantly a crystalline structure and exhibits higher thermal stability than hemicellulose. Cellulose decomposes in the temperature range between 315°C and 400°C [Stefanidis *et al.*, 2014; Yang *et al.*, 2007]. Finally, the third macrobiopolymer, lignin, is a polymer made out of aromatic moieties. In fact, lignin is the polymerization product of three different phenylpropanoid monomers (also known as monolignols): p-coumaryl alcohol, syringol and conferyl alcohol [Vanholme *et al.*, 2013]. These phenylpropanoid monomers have different functional groups (i.e. alcohol moieties) and double bonds, which gives rise to a wide array of chemical bond types (like ether bonds and covalent carbon-carbon bonds) by which these monolignols arrange themselves in the polymeric structure of lignin. Because of the range of chemical bond types, and each bond having a different activation energy for bond dissociation [Zakzeski *et al.*, 2010; Kim *et al.*, 2011], the thermal dissociation of lignin occurs over a very broad temperature range. Lignin can already start to decompose at temperatures above 160°C, but complete conversion may be reached at temperatures as high as 900–1000°C [Zhou *et al.*, 2013].

In pyrolysis, the reactions wherein the macrobiopolymers are broken down, with the accompanying release of gases and volatiles, are collectively known as the "primary pyrolysis reactions" and comprise depolymerization, fragmentation and rearrangement reactions [Di Blasi, 2008; Collard and Blin, 2014]. An overview of the general pyrolysis reaction scheme is given in Figure 2.1. The primary reactions not only release gases and vapors, they can directly give rise to char, which is often designated as "primary char" and consist of an aromatic polycyclic structure [Hajaligol *et al.*, 2001]. These primary reactions are highly endothermic [Antal and Grønli, 2003; Yang *et al.*, 2007]. The vapors, which consist of oxygenated compounds, are not stable in the vapor phase and tend to undergo a set of reactions, collectively known as the "secondary pyrolysis reactions" (see Figure 2.1). These secondary reactions consist of cracking reactions, which convert volatiles into smaller, non-condensable gas molecules; and condensation and polymerization reactions. The latter reactions will give rise to additional char, termed "secondary char" at the expense of primary volatiles [Wei *et al.*, 2006; Collard and Blin, 2014]. It is thought that the secondary reactions are exothermic. Additionally, minerals present in the biomass and already formed char can act as a

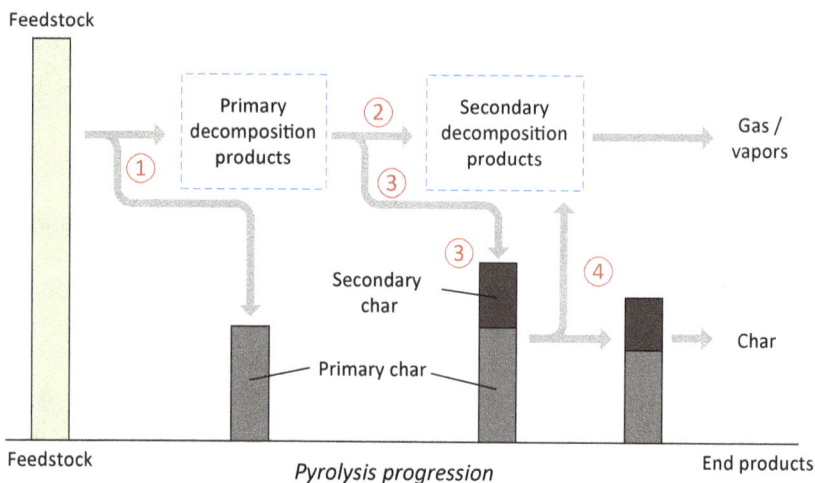

Figure 2.1: A generalized and simplified scheme of primary and secondary pyrolysis reactions occurring in lignocellulosic biomass. The bar length represents typical solid mass. Following reactions are indicated: (1) primary pyrolysis reaction; (2) secondary pyrolysis reactions (both homogeneous reactions in gas/vapor phase and heterogeneous reactions between solids and gas/vapor); (3) polymerization of intermediate volatiles; and (4) char maturation and outgassing.

catalyst to increase the rate of these secondary reactions and increase the yield of char [Ronsse *et al.*, 2012].

The grouping of pyrolysis reactions into primary and secondary also helps to explain the underlying mechanisms that differentiate fast and slow (as well as intermediate) pyrolysis. In fast pyrolysis, the yield of condensable vapors is maximized, which has to be achieved by suppressing the secondary reactions, as the latter would consume pyrolysis vapors to form additional gas and char. As such, short vapor residence times (max. 1–2 s) followed by quenching (condensing) are used to prevent the secondary reactions from happening [Venderbosch and Prins, 2010]. Because of the endothermic nature of primary pyrolysis reactions, fast pyrolysis requires heat to be added to the process, up to 1.8 MJ kg^{-1} of biomass [Venderbosch and Prins, 2010] and as there is no significant secondary char formation, the overall char yield of fast pyrolysis is low (15 wt.%). Slow pyrolysis, on the other hand, relies on process conditions which allow secondary reactions to take place to their full extent. As such, more secondary char is being formed, at the expense of condensable vapors. Another consequence of the extensive secondary reactions in slow pyrolysis is that the process is generally exothermic in nature [Park *et al.*, 2010; Manyà *et al.*, 2013].

2.3 Biochar Production Systems

From the above description of the dry thermochemical conversion processes, it is clear that a wide range of technologies in which char (biochar) is either obtained as the main product or as a co-product exist. The following sub-sections will give a brief overview of the different key reactor systems.

2.3.1 *Fast pyrolysis reactors*

Typical fast pyrolysis conditions include moderate temperatures (400–600°C), rapid heating rates (>100°C min^{-1}) combined with short residence times of both the biomass particles (0.5–2 s) [Demirbas, 2001] and vapor phase. To date, several reactor designs have been developed, tested and deployed to commercial scale. These fast pyrolysis reactors include entrained flow, ablative, vacuum moving bed, fluidized bed, auger and rotating cone reactors. Only the common reactor types will be discussed herein in more detail — for a more complete overview, the reader is referred to reviews by Venderbosch and Prins [2010], Bridgwater [2012] and Nachenius *et al.* [2013].

The most popular reactor design is the fluidized bed (see Figure 2.2(a)) in which a solid heat carrier, usually sand, is suspended in a stream of hot gas. If the upward gas velocity is set sufficiently high, the drag force upon the solid particles of the bed material overcomes the gravitational force, and hence the particles become suspended, or "fluidized". Into this bed of fluidized heat carrier particles, biomass, in the form of small particles, is injected. The vigorous motion of the fluidized heat carrier particles ensures optimal mixing behavior and rapid heat (and mass) transfer between the heat carrier, biomass particles and fluidizing gas. The char particles, formed out of the conversion of the biomass feedstock, typically have a lower density because of the devolatilization that takes place, as well as being of a smaller size as char particles tend to break apart because of their brittleness. Consequently, char particles are entrained from the

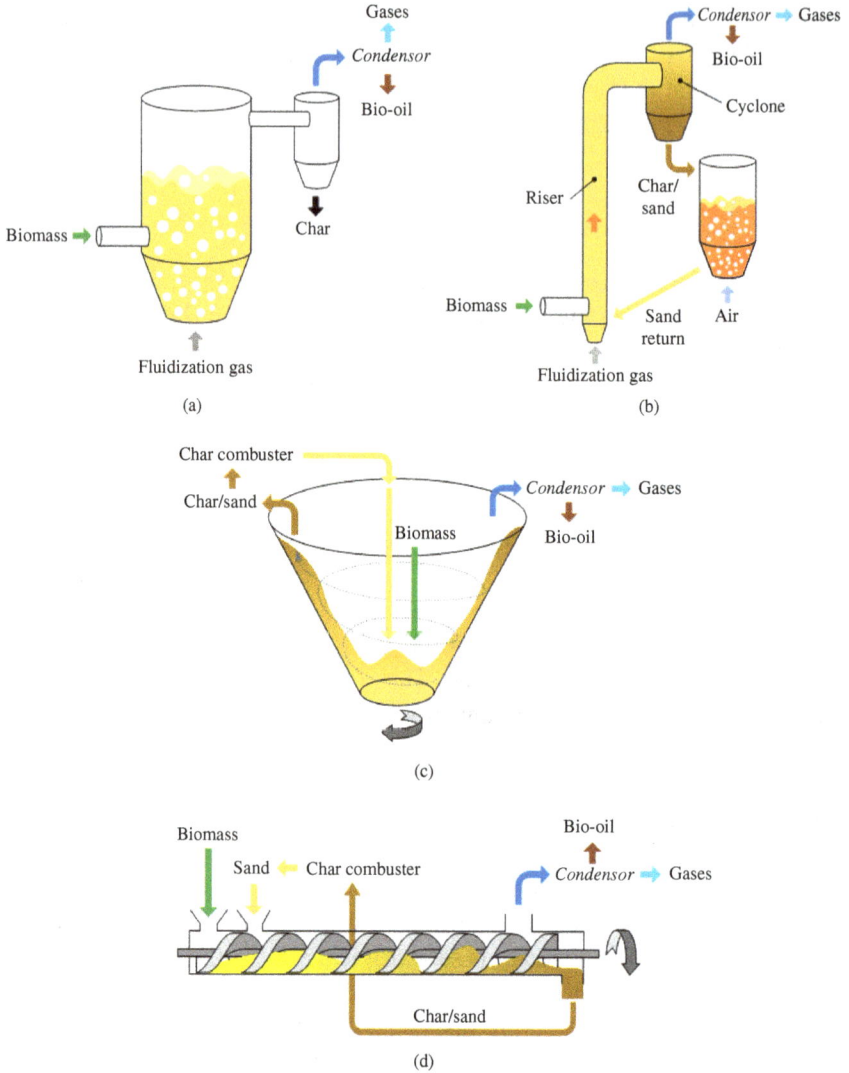

Figure 2.2:   Fast pyrolysis reactor types: (a) bubbling fluidized bed, (b) circulating fluidized bed, (c) rotating cone reactor and (d) screw or auger reactor.

reactor by the fluidizing gas (in which also the pyrolysis gas and vapors are mixed). Downstream of the reactor, a hot cyclone is employed to separate the char from the gas/vapor mixture, which subsequently undergoes condensation for the separation of bio-oil. The heat carrier particles (sand) do not leave the reactor, and this reactor type is better known as "bubbling fluidized bed".

If the fluidization gas velocity is increased even further, compared to the gas velocities used in bubbling fluidized beds, the heat carrier particles will also get entrained into the gas flow. Such a phenomena is used in a so-called "circulating" fluidized bed. Herein, the biomass/heat carrier mixture is lifted upwards through a narrow riser (i.e. turbulent fluidization regime). At the top of the riser, the vapors and gases are separated from the char/heat carrier mixture using a hot cyclone. Typically, the char is not separated from the heat carrier, but rather the mixture is processed in an air-blown (or oxygen-blown) fluidized bed where the char is combusted. By burning the char, the heat of combustion is absorbed by the heat carrier after which the heat carrier is fed to the bottom of the riser, thus closing the loop in this circulating fluidized bed. Although in practice all char is burned, to maintain stable energy supply (heat) only a fraction of the char is required and so these reactor configurations could still produce char, in theory. This remark is also valid for the rotating cone and auger reactors as displayed in Figure 2.2.

Other reactor designs include the rotating cone reactor, developed by Twente University (The Netherlands) [Wagenaar *et al.*, 1994] and Biomass Technology Group (The Netherlands) in which mechanical mixing of the heat carrier with the biomass particles is achieved by centrifugal forces generated in a rotating, inverted cone. Biomass particles and heat carrier are fed at the bottom of the cone and the mixture is spun upwards (Figure 2.2(c)). Screw or auger reactors (Figure 2.2(d)) consist of one or two, counter-rotating helical screws which transport a mixture of biomass and heat carrier through a cylindrical shell (barrel) or U-shaped trough. Screw reactors suffer from a minor penalty in bio-oil yields (typically 10% lower) due to longer vapor residence times (5–30 s) compared to fluidized bed fast pyrolysis reactors [Liaw *et al.*, 2012]. Additionally, screw reactors cannot be scaled up as much as fluidized bed reactors because of mechanical constraints. Both the rotating cone and the screw reactor can operate in the absence of an inert carrier gas, thereby yielding a highly concentrated vapor stream which can be handled by a fairly compact bio-oil condensation unit — the overall result is a reduction in cost. In practice though, both rotating cone and screw reactor have minimal consumption of inert carrier gas in order to increase the rate at which pyrolysis vapors are expelled out of the reactor as otherwise secondary pyrolysis reactions may become too extensive.

2.3.2 *Slow and intermediate pyrolysis reactors*

Many designs of slow pyrolysis units are based on or derived from designs used extensively up until the beginning of the 20th century for dry distillation of wood to produce chemicals as well as charcoal [Domac *et al.*, 2008]. As shown in Table 2.1, slow and intermediate pyrolysis units deploy lower heating rates compared to fast pyrolysis, and therefore the designs are not constrained by the need to use small biomass particles and rapid heat transfer from a reactor wall or a heat carrier into biomass particles. As a result, a wide range of batch and continuous pyrolysis units have been developed, but a full overview is outside of the scope of this chapter. Only selected key designs will be briefly discussed.

2.3.2.1 *Screw/auger pyrolysis units*

Screw pyrolysis units were already briefly introduced in Section 2.3.1 as these units can be operated as fast pyrolysers with solid heat carriers ensuring fast heat transfer into biomass. When used for slow or intermediate pyrolysis, heat is supplied externally by heating the reactor wall, or in some cases the screw, which subsequently heats the biomass transported by the screw from the feeding end towards the discharge where biochar is separated from the gases and vapors. Screw pyrolysis units are attractive for small- to medium-scale applications due to their relative simplicity, flexibility and ease of operation. This technology is however not suitable for large-scale units due to mechanical limitations stemming from operation of a moving metal screw inside a reactor at a high temperature. Current examples of this technology are the units developed and commercialized by Pyreg GmbH (www.pyreg.de).

2.3.2.2 *Drum and rotary kiln units*

Both drum and rotary kiln pyrolysis units consist of a horizontal or slightly inclined externally heated cylindrical shell through which biomass flows thanks to the action of paddles (drum kiln) or gravity and rotation of the shell (rotary kiln), see Figures 2.3(a) and (b), respectively. The residence time of biomass and biochar in this type of units can be controlled by speed of rotation and incline of the unit axis, and typically ranges from several minutes to several tens of minutes. Operating temperatures typically vary between 300°C and 900°C, spanning the range

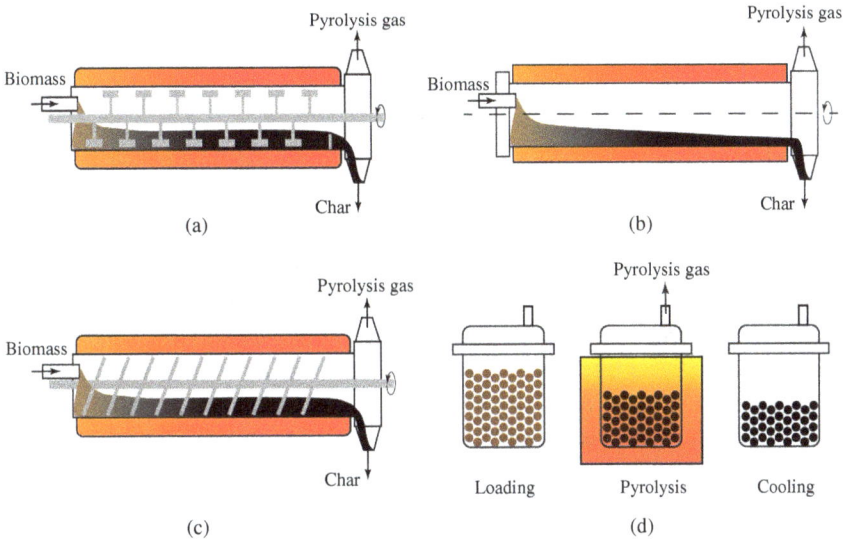

Figure 2.3:   Slow pyrolysis reactor types: (a) drum kiln, (b) rotary kiln, (c) screw pyrolyzer, and (d) batch pyrolysis retort.

from torrefaction to high temperature carbonization and even char activation. Heating is provided externally by heating the shell using gas or liquid fuels (including the co-products of pyrolysis). The heat is then transferred from the heated wall into particles constantly moving inside the reactor as they proceed towards the discharge. The advantage of these units is their feedstock flexibility in terms of composition and also particle size, as well as their scalability, especially for rotary kilns with no moving internals exposed to high temperatures. Rotary kiln technology is therefore suitable for units with capacities of few hundred tons per year up to over 100,000 t yr^{-1}.

2.3.2.3 *Batch units*

Compared to the above-mentioned examples of continuous pyrolysis technologies, batch pyrolysis units tend to be simpler and smaller, with capacities from under 100 kg to a few tons per batch. Biomass particles in the batch remain stationary during the whole process (except for movement caused by collapse of the batch with progress of pyrolysis). Unit designs can range from traditional mound or brick kilns used for centuries around the world for charcoal production to steel kilns with single or multiple pyrolysis chambers allowing semi-continuous operation. Typical operation of a batch unit consists of (a) loading of biomass feedstock, depending on the type of unit, this can be restricted to only certain types of biomass, such as stem wood; (b) carbonization, where the unit is ignited (part of the biomass feedstock is combusted, either internally or externally) and carbonization is allowed to complete; (c) emptying of the unit, recovering produced charcoal. Due to the nature of this operation, processing times are much longer than in continuous pyrolysis units, typically in the order of hours or even tens of hours, and it is often also much more labor intensive, limiting applicability to places with low labor cost. Furthermore, these units do not usually allow for a precise control of the pyrolysis conditions, which in some cases can result in formation of hot or cold spots (zones with too low or too high temperature), and subsequently in nonuniform and inconsistent quality of biochar. Some batch pyrolysis designs can be connected up together to enable semi-continuous operation where a number of units are operated simultaneously, each in a different stage of the pyrolysis cycle.

2.4 Relevant Biochar Properties

One of the biggest challenges for the biochar community is to understand how the operating conditions of the pyrolysis process can affect the properties of the produced biochar. This knowledge is crucial in order to produce engineered biochars for a given application. Despite the fact that pyrolysis of biomass is a relatively well-known technology, the yields and properties of products (including those of biochar) can be significantly affected by the operating conditions (e.g. peak temperature, heating rate, pressure, residence time of the vapor phase, particle size, and pyrolysis environment). Furthermore, the nature of the biomass source can also play a crucial role in determining the yield and properties of produced biochar [Manyà, 2012].

Depending on the further uses of the produced biochar, the key physicochemical properties to be monitored to assess its potential suitability can be different. Based on the current state of knowledge, the apparent key properties of biochar (and the analytical methods to assess them) as a function of its further use are summarized in the following subsections. Since most of the previous studies conducted in this field are based on biochars produced through slow pyrolysis, the information given below mainly refers to this thermochemical conversion process.

2.4.1 *Long-term environmental stability of biochar*

Regardless of the specific application of the produced biochar, it becomes highly interesting to analyze the potential stability of the carbon contained in the biochar (i.e. its recalcitrance or resistance to abiotic and biotic degradation) [Zimmerman, 2010]. Obviously, assessing the potential stability of biochar is especially relevant when biochar is finally applied into soil, since it can serve as an approximate indicator of the carbon sequestration potential of biochar. Nevertheless, quantifying the carbon stability of biochars in soil represents a great challenge, because carbon contained in biochar mineralizes more slowly than most other carbon sources [Singh *et al.*, 2012].

Given the fact that long-term (e.g. 5 years) experimental trials (e.g. laboratory incubation studies) are needed for estimating realistic values of carbon loss, a more practical solution is required [McBeath *et al.*, 2014]. In the last few years, several approaches to assess the potential stability of biochar have been proposed (see Table 2.2 for a summary). Among them, it should be highlighted that the approaches based on: (i) chemical and structural properties of biochar (e.g. percentage of aromatic carbon, deduced from solid-state [13]C Nuclear Magnetic Resonance), (ii) relative thermal stability of a given biochar to that of graphite (e.g. the R_{50} index proposed by Harvey *et al.* [2012]), and (iii) resistance to a chemical oxidant (e.g. H_2O_2, as proposed in the Carbon Stability Tool of the University of Edinburgh [Cross and Sohi, 2013]).

McBeath *et al.* [2015] observed a trade-off between reduced biochar yield and increased SPAC (stable polycyclic aromatic carbon) content with increasing pyrolysis temperature (in slow pyrolysis processes). These authors suggested that pyrolysis peak temperatures between 500°C and 700°C can be enough to maximize the SPAC content of produced biochars for most biomass sources. However, McBeath *et al.* [2015] also pointed out that a significant pool of aromatic biochar carbon exists in a "semi-labile" form that may not be stable over a long time (e.g. hundreds of years). In line with this, De la Rosa *et al.* [2018] recently observed an increase in the relative abundance of O-alkyl C and alkyl C at the expense of aromatic C in aged biochars (after 24 months of field aging). Thus, it should be kept in mind that produced biochars with high aromatic carbon contents do not necessarily result in highly stable biochars.

As noted in previous studies [Nguyen *et al.*, 2010; McBeath *et al.*, 2015; Azuara *et al.*, 2016], pyrolysis peak temperature seems to be the

Table 2.2:   Summary of different approaches that have been proposed to estimate the potential stability of biochar.

References	Approach	Validation
Singh *et al.* [2012]	The long-term stability of biochar can be predicted by determining the **proportion of nonaromatic C and the degree of aromatic C condensation**. These structural properties of biochar can be quantified from the solid-state ^{13}C Nuclear Magnetic Resonance (NMR) spectra.	A strong direct correlation between the structural parameters determined from the ^{13}C NMR spectra and the mean residence time (MRT) of the C in biochar was found. The MRT of 11 biochars, produced from different feedstocks at different temperatures, was determined from a 5-year incubation experiment with measurement of C mineralization rate (released CO_2).
Harvey *et al.* [2012]	A **recalcitrance index (R_{50})** was proposed as an indicator of the potential stability of biochar. This index can be determined from the thermogravimetric (TGA) curves of biochars under an atmosphere of air. The R_{50} index is the ratio between the temperature corresponding to 50% of mass loss measured for a given biochar and that corresponding to a reference material (graphite).	A mineralization study was conducted for 15 different biochars (produced from different biomass sources). From analyzing the data for total biochar-C loss after 1 year, a strong negative exponential-type relationship between biochar degradability and thermal recalcitrance (R_{50}) was observed.
Cross and Sohi [2013]	An accelerated ageing method, which reflects the oxidative nature of biochar degradation in soil, was proposed as a practical approach for screening the relative long-term stability of biochar. This method consists in determining the **carbon loss** of a given biochar **after oxidation with H_2O_2** at 80°C.	The proposed method was applied to a systematic set of biochar samples produced from five different biomass sources. The stability of carbon in these samples was 41.6–76.1%, loosely correlating with biochar molar O:C ratio. Thus, the method is able to capture the physical constraints to degradation that are not reflected in elemental ratios (e.g. the micromorphology of biochar arising from the specific feedstock, and the composition and physical distribution of its inorganic constituents).

most influential operating parameter in determining the potential stability of biochar. However, other operating factors (such as the mineral constituents of the biomass source and particle size) can also affect biochar stability. For instance, Manyà *et al.* [2016] reported excellent R_{50} and *stable-C*[1] values (0.635 and 95%, respectively) when a high-ash content refuse-derived fuel was previously added to the biomass feedstock (two-phase olive mill waste) at the ratio of 10 wt.% and subsequently pyrolyzed at a peak temperature of 600°C. In addition, Li *et al.* [2014] observed an improvement in the biological stability (in terms of microbial mineralization rate) when calcite ($CaCO_3$) or calcium dihydrogen phosphate ($Ca(H_2PO_4)_2$) was added to rice straw at the ratio of 20 wt.% and then pyrolyzed at 500°C. In an interesting study on the stability of rice straw-derived biochars, Guo and Chen [2014] suggested that the interaction between silicon (highly available in rice straw) and carbon can lead to a silica-encapsulated carbon structure, which can protect the biochar against physical and chemical oxidation.

Manyà *et al.* [2016] also observed a strong correlation between the R_{50} and *stable-C* variables, which were uncorrelated or weakly-correlated with other variables such us the molar H:C ratio and fixed-C content. In other words, predicting biochar stability only based on indicators derived from proximate and ultimate analyses cannot be sensitive enough.

In summary, both the Harvey's recalcitrance index and the University of Edinburgh's C stability tool appear as reasonably sensitive indicators of the potential stability of biochar, since they are able to capture the effects of a change in the operating parameters (not only in the peak temperature) of the slow pyrolysis process. In light of the current state of knowledge, the potential stability of biochar can be increased by increasing the peak temperature (at least up to 500°C, depending on the type of biomass source) and by the addition of some specific inorganic species. Further research is needed to better understand the role of these inorganic constituents and their interactions with the carbon matrix.

2.4.2 *Textural properties: Surface area and porosity*

Specific surface area (SSA) is closely related to a number of other biochar properties (such as water holding capacity and cation exchange capacity

---

[1] *Percentage of carbon retained after oxidation with H_2O_2 (Carbon Stability Tool).*

(CEC)). Therefore, biochars having a relatively large SSA could be required for a number of biochar applications [Weber and Quicker, 2018]. A certain porosity development can be expected as a result of the release of volatiles during the pyrolysis process. However and especially in the case of slow pyrolysis, this porosity is mainly constituted by narrow micropores (i.e. pore sizes below 1 nm), since pyrolysis peak temperature (up to 700°C) is not high enough to obtain a carbon material with a well-developed porosity and a hierarchical pore structure.

The SSA is usually estimated from the adsorption isotherm of N_2 at −196°C and adopting the BET model (Brunauer, Emmet, and Teller). Nevertheless, at cryogenic temperatures, the diffusion rate of the N_2 molecules into narrow micropores is extremely slow [Kim *et al.*, 2016], and thus the value of SSA determined from the N_2 adsorption isotherm might be underestimated. As an alternative, data from the adsorption isotherm of CO_2 at 0°C may be used to estimate the SSA values. In this sense, Kim *et al.* reported that the SSA values estimated using the BET model from CO_2 adsorption at 0°C can be considered as reasonable proxies for highly ultra-microporous materials.

The physical structure of biochar significantly changes during the course of biomass pyrolysis. As the pyrolysis temperature increases, especially above 400°C, a gradual increase in the SSA can be expected due to the promoted release of volatiles, which favors the development of some porosity, mainly in the ultra-micropore range [Chen *et al.*, 2017]. SSA, however, can begin to plateau at temperatures above 600°C or even decrease at higher temperatures (i.e. above 800°C) [Newalkar *et al.*, 2014]. The structural ordering and/or the coalescence of neighboring pores can explain this fact. Furthermore, the occlusion of the micropores due to the deposition of secondary char can also partly explain the decrease in SSA [Burhenne *et al.*, 2013].

In addition to the process temperature, pressure and heating rate can also affect the porosity development of biochar. Regarding the effect of pressure, Newalkar *et al.* [2014] observed a decrease in the CO_2-based SSA when pressure increased from 0.5 to 1.0 MPa (526 and 340 m^2 g^{-1}, respectively) during the fast pyrolysis of pine wood in an entrained flow reactor at a highest temperature of 600°C. However, increasing the pressure from 1.0 to 2.0 MPa led to an increased value of SSA (544 m^2 g^{-1}). This behavior can be related to the morphological evolution of biochar particles, which became more rounded and exhibited some bubbles in the surface when pressure was increased. Nevertheless, the extent of swelling remained constant at pressures above 1.0 MPa, whereas the driving force for the trapped volatiles to escape gradually increased with pressure, leading, as a result, to higher SSA at pressures above 1.0 MPa.

Recently, Manyà *et al.* [2018] reported an almost constant CO_2-based SSA of 240–270 m^2 g^{-1} for vine shoots-derived biochars, which were produced through atmospheric or pressurized (1.0 MPa) slow pyrolysis at 600°C, and using N_2 or CO_2 as carrier gas. These results seem to be in contradiction with those reported by Newalkar *et al.* [2014]. However, it should be noted that the reactor configuration could play a key role in the textural properties of produced biochars. In the work of Manyà *et al.* [2018], relatively large biomass particles were pyrolyzed at very low heating rates in a packed-bed reactor, leading to relatively high contact times between the volatiles and the carbonaceous solid. As mentioned above, the promotion of secondary charring reactions can result in a decrease in SSA because of the deposition of secondary char and the subsequent pore blockage. On the other hand, using a pyrolysis atmosphere composed of CO_2 and N_2 (60/40 vol.%) — instead of pure N_2 — at atmospheric pressure and 500°C resulted in a marked increase in the SSA of the resulting wheat-straw derived biochar (from 200 to 380 m^2 g^{-1}), which was also produced in a packed-bed slow pyrolysis system [Greco *et al.*, 2018]. Nonetheless, the magnitude of the effect of CO_2 on porosity development seems to be dependent on, among other factors, the biomass feedstock (see Figure 2.4, where the CO_2 adsorption isotherms at 0°C for several wheat straw- and vine shoots-derived biochars are shown).

2.4.3 *Surface functional groups and related properties*

It is generally accepted that the availability of functional groups on the surface of biochar decreases with increasing the process temperature. Luo *et al.* [2015] reported a marked increase in the contents of surface acidic functional groups (carboxylic, lactonic and phenolic; which were measured by Boehm titration) as pyrolysis temperature increased from 300 to 500°C for different biomass sources. The removal of oxygenated functional groups leads to the formation of negatively charged surfaces and a decrease in the CEC [Suliman *et al.*, 2016]. Nevertheless, the dependence of CEC as a function of the pyrolysis temperature could depend on the biomass feedstock. In this sense, Suliman *et al.* [2016] observed that, for biochars produced from Douglas Fir wood, the CEC was kept almost

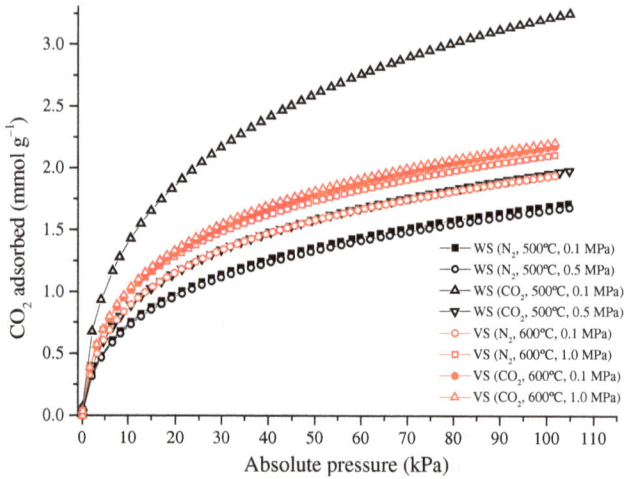

Figure 2.4: CO_2 adsorption isotherms at 0°C for several biochars derived from wheat straw (WS) and vine shoots (VS). The captions in the legend blocks indicate: the type of pyrolysis atmosphere (pure N_2 or a CO_2-containing mixture), highest pyrolysis temperature and absolute pressure.

Sources: Data from Greco *et al.* [2018]; Manyà *et al.* [2018].

constant regardless of the pyrolysis temperature (in the range of 350–550°C). One possible explanation for this behavior was the relatively high content of surface carboxylic groups in the Douglas Fir wood-derived biochars (even at 550°C).

The pH of biochar in water is usually alkaline and generally increases with the pyrolysis temperature. The value of pH, however, also depends on the availability of basic functional groups, which are mainly associated to the inorganic matter fraction [Suliman *et al.*, 2016]. A similar trend is usually observed for the electric conductivity (measured in water), which gradually increases (to a higher or lesser extent depending on the feedstock) with the pyrolysis temperature. This fact can be related to the progressive loss of acidic functional groups and the gradual increase in the ash content as the pyrolysis temperature rises.

In addition, data obtained from FTIR (Fourier transform infrared spectroscopy) revealed a gradual decrease in the stretching of the hydrogen-bonded OH group (hydroxyl, or alcohol/phenolic — OH groups) when the pyrolysis highest temperature increased from 200 to 500°C [Chen *et al.*, 2017]. Some bands (e.g. those attributed to aliphatic C and carboxylic C=O) started to disappear at temperatures above 500°C.

Conversely, the aromatic stretching can remain unchanged or even increase when the temperature rises, since biochar becomes progressively more aromatic.

References

Antal, M. J. and Grønli, M. (2003). The art, science, and technology of charcoal production. *Ind. Eng. Chem. Res.*, 42(8), pp. 1619–1640.

Azuara, M., Baguer, B., Villacampa, J. I., Hedin, N. and Manyà, J. J. (2016). Influence of pressure and temperature on key physicochemical properties of Corn Stover-derived biochar. *Fuel*, 186, pp. 525–533.

Bain, R. and Broer, K. (2011). Gasification, In: Brown, R. C. (ed.). *Thermochemical processing of biomass — conversion into fuels, chemicals and power* (John Wiley and Sons), pp. 47–77.

Balat, M., Balat, M., Kirtay, E. and Balat, H. (2009). Main routes for the thermoconversion of biomass into fuels and chemicals. Part 1: Pyrolysis systems, *Energ. Convers. Manage.*, 50, pp. 3147–3157.

Basu, P. (2013). *Biomass Gasification, Pyrolysis and Torrefaction: Practical Design and Theory*, 2nd Ed. (Academic Press, Burlington, USA).

Bridgwater, A. V. (2012). Review of fast pyrolysis of biomass and product upgrading, *Biomass Bioenergy*, 38, pp. 68–94.

Bridgwater, A. V. and Peacocke, G. V. C. (2000). Fast pyrolysis processes for biomass. *Renew. Sustain. Energy Rev.*, 4, pp. 1–73.

Burhenne, L., Damiani, M. and Aicher, T. (2013). Effect of feedstock water content and pyrolysis temperature on the structure and reactivity of spruce wood char produced in fixed bed pyrolysis. *Fuel*, 107, pp. 836–847.

Chen, Y., Zhang, X., Chen, W., Yang, H. and Chen, H. (2017). The structure evolution of biochar from biomass pyrolysis and its correlation with gas pollutant adsorption performance. *Bioresour. Technol.*, 246, pp. 101–109.

Collard, F. X. and Blin, J. (2014). A review on pyrolysis of biomass constituents: Mechanisms and composition of the products obtained from the conversion of cellulose, hemicelluloses and lignin, *Renew. Sustain. Energy Rev.*, 38, pp. 594–608.

Cross, A. and Sohi, S. P. (2013). A method for screening the relative long-term stability of biochar. *GCB Bioenergy*, 5, pp. 215–220.

Czernik, S. and Bridgwater, A. V. (2004). Overview of applications of biomass fast pyrolysis oil. *Energy Fuels*, 18, pp. 590–598.

De la Rosa, J. M., Rosado, M., Paneque, M., Miller, A. Z. and Knicker, H. (2018). Effects of aging under field conditions on biochar structure and composition: Implications for biochar stability in soils. *Sci. Total Environ.*, 613–614, pp. 969–976.

Demirbas, A. (2001). Biomass resource facilities and biomass conversion processing for fuels and chemicals. *Energy Convers. Manage.*, 42, pp. 1357–1378.

Di Blasi, C. (2008). Modeling chemical and physical processes of wood and biomass pyrolysis. *Prog. Energy Combust.*, 34, pp. 47–90.

Domac, J., Trossero, M. and Siemons, R. (2008). *Industrial Charcoal Production*, TCP/CRO/1301 (A) Development of a sustainable charcoal industry (FAO, Zagreb, Croatia).

Greco, G., Videgain, M., Di Stasi, C., González, B. and Manyà, J. J. (2018). Evolution of the mass-loss rate during atmospheric and pressurized slow pyrolysis of wheat straw in a bench-scale reactor. *J. Anal. Appl. Pyrolysi*, 136, pp. 18–26.

Grønli, M. and Melaaen, M. C. (2000). Mathematical model for wood pyrolysis — comparison of experimental measurements with model predictions. *Energy Fuels*, 14, pp. 791–800.

Guo, J. and Chen, B. (2014). Insights on the molecular mechanism for the recalcitrance of biochars: Interactive effects of carbon and silicon components. *Environ. Sci. Technol.*, 48, pp. 9103–9112.

Hajaligol, M., Waymack, B. and Kellogg, D. (2001). Low temperature formation of aromatic hydrocarbon from pyrolysis of cellulosic materials, *Fuel*, 80(12), pp. 1799–1807.

Harvey, O. R., Kuo, L. J., Zimmerman, A. R., Louchouarn, P., Amonette, J. E. and Herbert, B. E. (2012). An index-based approach to assessing recalcitrance and soil carbon sequestration potential of engineered black carbons (Biochars). *Environ. Sci. Technol.*, 46, pp. 1415–1421.

Isahak, W. N. R. W, Hisham, M. W. M. Yarm, M. A. and Hin, T. Y. Y. (2012). A review on bio-oil production from biomass using pyrolysis method, *Renew. Sustain. Energy Rev.*, 16, pp. 5910–5923.

Kim, S., Chmely, S. C., Nimlos, M. R., Bomble, Y. J., Foust, T. D., Paton, R. S. and Beckham, G. T. (2011). Computational study of bond dissociation enthalpies for a large range of native and modified lignins, *J. Phys. Chem. Lett.*, 2, pp. 2846–2852.

Kim, K. C., Yoon, T. U. and Bae, Y. S. (2016). Applicability of using CO_2 adsorption isotherms to determine BET surface areas of microporous materials. *Microporous Mesoporous Mater.*, 224, pp. 294–301.

Kruse, A. and Dahmen, N. (2015). Water — A magic solvent for biomass conversion, *J. Supercrit. Fluid.*, 96, pp. 36–45.

Lehmann, J. and Joseph, S. (2009). Biochar for environmental management: An introduction. In: Lehmann, J. and Joseph, S. (eds.). *Biochar for Environmental Management: Science and Technology* (Earthscan, London), pp. 1–12.

Li, F., Cao, X., Zhao, L., Wang, J. and Ding, Z. (2014). Effects of mineral additives on biochar formation: Carbon retention, stability, and properties. *Environ. Sci. Technol.*, 48, pp. 11211–11217.

Liaw, S. S., Wang, Z., Ndegwa, P., Frear, C., Hu, S., Li, C. Z. and Garcia-Perez, M. (2012). Effect of pyrolysis temperature on the yield and properties of bio-oils obtained from the auger pyrolysis of Douglas-fir wood, *J. Anal. Appl. Pyrolysis*, 93, pp. 52–62.

Lu, Q., Li, W. Z. and Zhu, X. F. (2009). Overview of fuel properties of biomass fast pyrolysis oils. *Energy Convers. Manage.*, 50, pp. 1376–1383.

Luo, L., Xu, C., Chen, Z. and Zhang, S. (2015). Properties of biomass-derived biochars: Combined effects of operating conditions and biomass types. *Bioresour. Technol.*, 192, pp. 83–89.

Manyà, J. J. (2012). Pyrolysis for biochar purposes: A review to establish current knowledge gaps and research needs. *Environ. Sci. Technol.*, 46, pp. 7939–7954.

Manyà, J. J., Laguarta, S. and Ortigosa, M. A. (2013). Study on the biochar yield and heat required during pyrolysis of two-phase olive mill waste. *Energy Fuels*, 27(10), pp. 5931–5939.

Manyà, J. J., Alvira, D., Azuara, M., Bernin, D. and Hedin, N. (2016). Effects of pressure and the addition of a rejected material from municipal waste composting on the pyrolysis of two-phase olive mill waste. *Energy Fuels*, 30, pp. 8055–8064.

Manyà, J. J., González, B., Azuara, M. and Arner, G. (2018). Ultra-microporous adsorbents prepared from vine shoots-derived biochar with high CO_2 uptake and CO_2/N_2 selectivity. *Chem. Eng. J.*, 345, pp. 631–639.

Masek, O., Brownsort, P., Cross, A. and Sohi, S. (2013). Influence of production conditions on the yield and environmental stability of biochar. *Fuel*, 103, pp. 151–155.

McBeath, A. V., Smernik, R. J., Krull, E. S. and Lehmann, J. (2014). The influence of feedstock and production temperature on biochar carbon chemistry: A solid-state 13C NMR study. *Biomass Bioenergy*, 60, pp. 121–129.

McBeath, A. V., Wurster, C. M. and Bird, M. I. (2015). Influence of feedstock properties and pyrolysis conditions on biochar carbon stability as determined by hydrogen pyrolysis. *Biomass Bioenergy*, 73, pp. 155–173.

McKendry, P. (2002). Energy production from biomass (part 3): Gasification technologies. *Bioresour. Technol.*, 83, pp. 55–63.

Mohan, D., Pittman, C. U. Jr. and Steele, P. H. (2006). Pyrolysis of wood/biomass for bio-oil: A critical review. *Energy Fuels*, 20, pp. 848–889.

Nachenius, R. W., Ronsse, F., Venderbosch, R. H. and Prins, W. (2013) Biomass pyrolysis. In: Murzin, D. Y. (ed). *Adv. Chem. Eng.*, 42 (Academic Press, Burlington), pp. 75–139.

Newalkar, G., Lisa, K., Damico, A. D., Sievers, C. and Agrawal, P. (2014). Effect of temperature, pressure, and residence time on pyrolysis of pine in an entrained flow reactor. *Energy Fuels*, 28, pp. 5144–5157.

Nguyen, B. T., Lehmann, J., Hockaday, W. C., Joseph, S. and Masiello, C. A. (2010). Temperature sensitivity of black carbon decomposition and oxidation. *Environ. Sci. Technol.*, 44, pp. 3324–3331.

Park, W. C., Atreya, A. and Baum, H. R. (2010). Experimental and theoretical investigation of heat and mass transfer processes during wood pyrolysis, *Combust. Flame*, 157(3), pp. 481–494.

Qian, K. Z., Kumar, A., Shang, H. L., Bellmer, D. and Huhnke, R. (2015). Recent advances in utilization of biochar, *Renew. Sustain. Energy Rev.*, 42, pp. 1055–1064.

Ronsse, F., Bai, X., Prins, W. and Brown, R. C. (2012). Secondary reactions of levoglucosan and char in the fast pyrolysis of cellulose, *Environ. Prog. Sustain.*, 31, pp. 256–260.

Saha, B. (2003). Hemicellulose bioconversion, *J. Ind. Microbiol. Biot.*, 30, pp. 279–291.

Singh, B. P., Cowie, A. L. and Smernik, R. J. (2012). Biochar carbon stability in a clayey soil as a function of feedstock and pyrolysis temperature. *Environ. Sci. Technol.*, 46, pp. 11770–11778.

Stefanidis, S. D., Kalogiannis, K. G., Iliopoulou, E. F., Michailof, C. M., Pilavachi, P. A. and Lappas, A. A. (2014). A study of lignocellulosic biomass pyrolysis via the pyrolysis of cellulose, hemicellulose and lignin, *J. Anal. Appl. Pyrolysis*, 105, pp. 143–150.

Suliman, W., Harsh, J. B., Abu-Lail, N. I., Fortuna, A. M., Dallmeyer, I. and Garcia-Perez, M. (2016). Influence of feedstock source and pyrolysis temperature on biochar bulk and surface properties. *Biomass Bioenergy*, 84, pp. 37–48.

Van der Stelt, M. J. C., Gerhauser, H., Kiel, J. H. A. and Ptasinski, K. J. (2011). Biomass upgrading by torrefaction for the production of biofuels: A review. *Biomass Bioenergy*, 35, pp. 3748–3762.

Vanholme, B., Desmet, T., Ronsse, F., Rabaey, K., Van Breusegem, F., De Mey, M., Soetaert, W. and Boerjan, W. (2013). Towards a carbon-negative sustainable bio-economy, *Front. Plant Sci.*, 4, p. 174.

Venderbosch, R. H. and Prins, W. (2010). Fast pyrolysis technology development, *Biofuels, Bioprod. Biorefin.*, 4, pp. 178–208.

Wagenaar, B. M., Prins, W. and Van Swaaij, W. P. M. (1994). Pyrolysis of biomass in the rotating cone reactor: Modelling and experimental justification, *Chem. Eng. Sci.*, 49, pp. 5109–5126.

Weber, K. and Quicker, P. (2018). Properties of biochar. *Fuel*, 217, pp. 240–261.

Wei, L., Xu, S., Zhang, L., Zhang, H., Liu, C., Zhu, H. and Lio, S. (2006). Characteristics of fast pyrolysis of biomass in a free fall reactor. *Fuel Process. Technol.*, 87(10), pp. 863–871.

White, J. E., Catallo, W. J. and Legendre, B. L. (2011). Biomass pyrolysis kinetics: A comparative critical review with relevant agricultural residue case studies. *J. Anal. Appl. Pyrolysis*, 91, pp. 1–33.

Williams, P. T. and Besler, S. (1996). The influence of temperature and heating rate on the slow pyrolysis of biomass, *Renew. Energy*, 7, pp. 233–250.

Yang, H., Yan, R., Chen, H., Lee, D. H. and Zheng, C. (2007). Characteristics of hemicellulose, cellulose and lignin pyrolysis, *Fuel*, 86, pp. 1781–1788.

Yang, S. I., Wu, M. S. and Wu, C. Y. (2014). Application of biomass fast pyrolysis part I: Pyrolysis characteristics and products. *Energy*, 66, pp. 162–171.

Zakzeski, J., Bruijninck, P. C. A., Jongerius, A. L. and Weckhuysen, B. M. (2010). The catalytic valorization of lignin for the production of renewable chemicals. *Chem. Rev.*, 110, pp. 3552–3599.

Zimmerman, A. R. (2010). Abiotic and microbial oxidation of laboratory-produced black carbon (biochar). *Environ. Sci. Technol.*, 44, pp. 1295–1301.

Zhou, H., Long, Y. Q., Meng, A. H., Li, Q. H. and Zhang, Y. G. (2013). The pyrolysis simulation of five biomass species by hemi-cellulose, cellulose and lignin based on thermogravimetric curves. *Thermochim. Acta.*, 566, pp. 36–43.

Chapter 3

Water Holding Capacity of Biochar and Biochar-Amended Soils

Wenceslau Geraldes Teixeira[*],[§], Frank Verheijen[†]
and Jean Dalmo de Oliveira Marques[‡]

Brazilian Agricultural Research Corporation (EMBRAPA), Embrapa Solos, Brazilia DF, Brazil

†*Cesam, University of Aveiro, Aveiro, Portugal*

‡*Federal Institute of Education, Science and Technology of the Amazonas (IFAM), Campus Center of Manaus, Manaus, Brazil*

§*wenceslau.teixeira@embrapa.br*

Abstract

Water holding capacity (WHC) is defined as the ability of a medium to hold water controlled by its geometry and composition. Indication of methods to evaluate WHC of biochar and soil mixed with biochar are outlined. Studies involving the use of biochar applications and effects on WHC are reviewed and discussed. The feasibility of adding biochar to soil to enhance water holding capacity is still controversial not only because of non-significant effects in many soils, but also because the feedstock and the pyrolysis technique used to produce biochar and the consequences in WHC are still not completely understood. Biochar additions may be more effective to enhance plant-available water (PAW) in soil with a coarse texture. Caution is necessary in application of large

61

amounts of biochar produced from not well-studied feedstocks or pyrolysis technique that is not produced to comply with the biochar standards or certificates, as some consequence may not be reversible.

3.1 Introduction

Water holding capacity (WHC) is defined as the ability of a medium to hold water controlled by its geometry and composition. WHC may be specified by the amount of water in the medium at specified suction potentials. Common values for potentials to measure WHC in soils are 0, 6, 10, 33, 1000 and 1500 kPa. The WHC is used to determine plant available water (PAW), which is calculated as the difference between WHC at field capacity (FC) minus WHC at equilibrium with 1500 kPa (Permanent Wilting Point — PWP). Specific requirements for evaluation of WHC of biochar samples and soil plus biochar mixtures were identified by Tammeorg *et al.* [2017]. A pioneer study of the effect of biochar on WHC in sandy, loamy and clayey soils was carried out by Tryon [1948]. It showed a small increase in sandy and a decrease in clayey soils. A commonly reported effect of biochar application to soil is an improvement in WHC [Baiamonte *et al.* 2019; Blanco-Canqui, 2017; Omondi *et al.*, 2016; Obia *et al.*, 2016; Sun and Lu, 2014; Abel *et al.*, 2013; Tammeorg *et al.*, 2013; Verheijen *et al.*, 2010; Sohi *et al.*, 2009]. However, some studies have also shown no effect of biochar addition or reductions in WHC at different potentials [Paneque *et al.*, 2016; Jeffery *et al.*, 2015; Ojeda *et al.*, 2015; Hardie *et al.*, 2014; Major *et al.*, 2012]. Different biochars vary in their properties to enhance WHC since their characteristics are controlled mainly by the thermal conversion technology and feedstock types [Grey *et al.*, 2015]. In this chapter, some methods and indications to evaluate WHC of biochar and soil mixed with biochar are outlined. Moreover, studies involving the use of biochar applications and effects on WHC are reviewed and discussed.

3.2 Evaluation of WHC of Soil, Biochar and Soil Mixed with Biochar

There are different methods to obtain WHC of biochar or soils mixed with biochar. The most common procedure is to equilibrate samples at a chosen potential and then gravimetrically determine the moisture. The moisture may be expressed gravimetrically or volumetrically and related with the selected potential. Normally, it is done in the laboratory. Details about evaluations in soil samples are available in Teixeira and Behring [2017] and Dane and Hoppmans [2002]. Field methods using water moisture and potential meters are also available [Romano and Santini, 2002]. Some countries have regulations about WHC for commercial use of growing media. In Brazil and other countries, it should be measured using a potential of 10 kPa. It is done in a very simple way by soaking the material for one to two days, followed by draining it over paper filter. The WHC is then calculated as the water amount (g) divided by the dry matter [Ahn *et al.*, 2008]. In the following section, some particularities of methods to evaluate WHC will be presented and discussed.

3.2.1 *Evaluation of water held at zero potential or total porosity*

The measurement of a saturated sample (0 kPa) is a way to estimate the total porosity of the sample. Normally, an undisturbed sample is collected by hammering or jacking in a steel cylinder with known volume inserted into the soil. These cylinders are covered by a piece of nylon secured with an elastic band. Samples are saturated slowly from bottom to top, allowing air in soil pores to escape and thereby reducing entrapped air in the sample. Wetting in this way, and leaving the sample saturated for 24–48h, also avoids heterogeneous soil wetting caused by hydrophobic parts of the soil or biochar. The same procedure can be used to evaluate disturbed samples. Cylinders are filled and packed with biochar, soil or a soil–biochar mixture, which are saturated, and then the difference of the volume between the cylinder volume and water volume, that is normally considered with a density of 1,00 g cm^{-3}, may be considered the total porosity or volumetric water at saturation. In soil samples, moisture at saturation may range from 0.90 to 0.30 cm^{-3} for peat or Histosols samples to very compact clayey soils (e.g. clayey Gleysols). Total porosity determined by this way can be upto 25% less than total porosity calculated using bulk and particle densities values due to entrapped air in the samples. Total porosity of soil mixed with biochar was reviewed by Blanco-Canqui [2017] and Omondi *et al.* [2016] who concluded that the addition of biochar enhance total porosity; in most studies the magnitude of the effect varies with the amount and type of biochar and the soil where it was applied.

3.2.2 *Evaluation of water held at 10–100 kPa*

Büchner funnels or sand tension tables can be used to equilibrate samples from potential 10 to 100 kPa [Teixeira and Behring, 2017; Dane and Hoppmans, 2002]. The principle of these methods is that the samples are placed in hydraulic contact with a medium whose pores are so small that they remain in a saturated state at these potential levels (i.e. the bubbling point of the material is higher than the potential measured). In the Büchner funnels and the sand box tension tables, the suction is created by a hanging water column. After the attainment of the equilibrium with the applied potential, which is observed by cessation of outflow drainage or regularly weighting until the weight is constant, samples are weighed then oven-dried (105°C during 24–48 hours) and weighed again. The mass of water and dry matter of the sample can be calculated. Because the Büchner funnel requires a separate apparatus for each sample, sand tension tables are more common in laboratories that deal with large number of samples. A tension table is a simple apparatus, easily operated at low cost. To reduce loss by evaporation, the tension table should be covered. For low potentials, the equilibrium may be achieved within 2–4 days.

3.2.3 *Evaluation of water held at 10–50 kPa*

The air-entry value of fine sand precludes the use of sand suction tables above 10 kPa. A mixture of sand and kaolinite creates a medium with high bubbling point allowing application of potential >10 kPa and smaller than 600 kPa. The selected potential is created by using a vacuum pump. The equilibrium time is also checked by regularly weighing and more time is needed to reach equilibrium (5–10 days).

3.2.4 *Evaluation of water held at 600–1500 kPa*

One of most popular devices in soil physics laboratories are the pressure plates called Richard chambers. Saturated ceramic plates are confined in a chamber and regulators and compressor gauges allow applying a positive pressure on the samples. Once the chamber has been sealed, a gas pressure is applied to the air space above the samples, and water moves downward from the samples through the ceramic and can be drained. To maintain the potential, the porous medium (water saturated ceramic plates) and samples are contained within a pressurized chamber while the underside of the porous medium is maintained at atmospheric pressure. The equilibrium is assessed when the outflow of water from the drain tubes ceases. Then the samples are removed, weighed and oven-dried and weighed again. The equilibrium time depends on the potential and the height of the sample, and ranges from 10 to 40 days. At high pressures, a sample height of 1–3 cm is more convenient. Because the water in samples at high pressures is held in small pores, it is acceptable to use disturbed samples.

3.2.5 *Potential > 1000 kPa–300 MPa — The dew point psychrometer*

Water potential is a measurement of the energy status of the water that indicates how tightly water is bound, structurally or chemically, within a substance. The most common device to measure water potential of a sample is psychrometer (WP4C, Meter, USA) that uses the chilled mirror dew point technique to measure the water potential of a disturbed sample. It has a high accuracy in the dry range and is used normally to measure the PWP (1.5 MPa), but it may measure potentials until 300 MPa. Choosing the methods to evaluate WHC of biochar or soil mixed with biochar will depend on the available equipment to check the desired potential, and time demands of the analyses. It is recommended to use undisturbed samples in field experiments to evaluate WHC in low suction range, since the structure of the soil affects the water retention in low potentials. For high potentials, it is better to use disturbed samples with low weight or the time of equilibrium will be very long. The time taken for characterization of many potentials can be 3–4 months. A summary of the methods to evaluated WHC are shown in Table 3.1.

Two limitations of laboratory methods in quantifying the effect of biochar amendment on soil WHC are possible changes in: (i) soil volume; and (ii) infiltration capacity.

(i) **Soil volume:** The decrease in soil bulk density caused by biochar amendment is associated with a soil volume increase. The extra water stored in this "extra soil" is relevant to soil-based ecosystem services [Kibblewhite *et al.*, 2007]. For example, increased rooting depth gives plants' roots access to more water, as well as nutrients. Most laboratory methods do not consider this additional soil water, thereby reducing the

Table 3.1:   Methods to evaluate WHC at different potentials on samples of soil, biochar or soil with biochar.

Method	Potential	Typical range	Cost of apparatus
Büchner funnel	Hanging water column	0–100	Very low
Sand tension tables	Hanging water column	0–100	Low
Kaolin-sand tension tables	Vacuum pump	100–500	Medium
Pressure plates (Richard chambers)	Compressors and gauges	100–1500	High
Psychrometry	Dew point	1000–30000	High

environmental representation of biochar effects on soil moisture. To increase environmental representation, laboratory studies using soil columns can include the additionally stored water expressed as "total soil water storage" by basing the columns on pre-amendment soil mass, thereby allowing greater soil volume for biochar-amended soil, rather than keeping all treatments at the same soil volume [Verheijen *et al.*, 2019b].

(ii) **Infiltration capacity:** WHC under field conditions depends on the infiltration capacity of the soil, as well as on what proportion of rainfall enters the soil rather than flows downhill along the soil surface, i.e. what the soil does not receive, it cannot hold onto. Biochar has been shown to either increase or decrease infiltration capacities of soils, with the latter mostly associated with sandy soils [Wang *et al.*, 2017]. In soils susceptible to crusting, biochar may cause a particularly large increase in infiltration capacity, depending on other soil conditions. For example, Abrol *et al.* [2016] measured a 1.7 times increased infiltration rate for a non-calcareous loamy sand, but no effect was found for a calcareous loamy soil. Considering these and other limitations of laboratory methods, more relevant information may be obtained in field studies.

3.3 Field Methods to Evaluate WHC

3.3.1 *Evaluating soil moisture in the field*

Gravimetric and volumetric water content of soil samples can be measured by sampling, drying, and weighing of soil samples using disturbed samples or core samples (e.g. Dane and Topp, 2002). Gravimetric soil moisture can be converted to volumetric soil moisture by using the corresponding bulk density value. Grossman and Reinsch [2002] give details for sampling and evaluating bulk density. Addition of biochar to soil tends to reduce bulk density. Reviews of biochar's effect on bulk density values are provided by Blanco-Canqui [2017] and Omondi *et al.* [2016].

Time-domain reflectometry (TDR) is a commonly used technique to measure volumetric soil moisture content (θ) using the travel time of electromagnetic waves. In frequency domain reflectometry (FDR), the calibration is done with the equilibrated frequency. Both are based on measurement of the soil bulk dielectric constant or soils' apparent permittivity (ε_a) and estimates θ using calibration curves. Kameyama *et al.* [2014] found TDR-based measurements to over-estimate SMCs when the soil contained biochar made at higher pyrolysis temperatures, possibly because of conductive and dielectric losses. They recommend the use of the real component (ε_r') of the soil's complex relative permittivity instead of ε_a for TDR measurements of SMC when high temperature pyrolysis biochar was added. Hartie *et al.* [2014] used manual TDR probes to evaluate the effect of biochar application in soil moisture (0–6 cm depth) and calculated the cumulative soil moisture content over time, finding no significant differences between the plots control and with biochar. Tammeorg *et al.* [2013] also measured soil moisture weekly and manually using TDR probes in different layers in a field experiment in Finland. Their results show a possible drought alleviation in plots where biochar was added, in which an increase of 2% of moisture corresponded to a 12 mm of additional water stored in the layer between 0 and 60 cm depth. In another field study using TDR probe to evaluate soil moisture, in Finland, comparing plots with e without biochar also found an increase in soil moisture in biochar plots, but it not corresponded to an increase in yield in winter wheat [Tammeorg *et al.*, 2014].

Castelline *et al.* [2019] used yield simulation model — Decision Support System for Agrotechnology Transfer (DSSAT) — to estimate the water balance and to forecast durum wheat yields, in a field with and without biochar applications. The results indicate that a moderate addition of biochar has the potential to increase the durum wheat yield. Indication can be gained from a study of biochar enriched soil in a Belgian maize field [Kerre *et al.*, 2017]. The authors found the PAW to be increased by 13% and observed the highest yield increase in the driest year. However, more information about the PAW and productivity for a range of biochar–soil–climate combinations needs to be validated with field measurements.

3.3.2 *Evaluating soil water matric potential in the field*

Tensiometers consist of three basic interconnected elements: a porous cup, a water reservoir and a measurement device (vacuum gauges to pressure transductors). Tensiometers are limited to appropriately work in the limited range from 0 to 80 kPa [Young and Sisson, 2002]. In biochar experiments, tensiometers were used by Steiner *et al.* [2002] to evaluate soil water matric potential in planting holes of banana plantation with and without addition of biochar, in yellow Ferralsol in the Central Amazon. They found very small differences in the soil water potential between the treatments and did not publish the data (personal information).

Automatic tensiometers (Models T8 — T4, Meter, USA) and sensors using other techniques such as the *Equitensiometer* — EQ3 that measures in the range from 0 to 1000 kPa (Delta T, UK) and *Teros — 21* (Meter, USA) that measures in the range 0 to 100 MPA are available, but biochar field experiments that monitor soil water potential are missing.

3.4 Estimation of FC, PWP and Plant-Available Water in Biochar and Soil Mixed with Biochar Samples

Field capacity (FC) is defined as the amount of water held in the soil after excess gravitational water has drained away and the rate of downward movement has materially decreased. Potential values for FC range from 6 to 33 kPa. As the movement of water downwards does not cease, FC is considered as a value that allows comparisons rather than its original physical concept [Romano and Santini, 2002]. Even though FC is not an exact value, the reason that the term has been brought back into the literature probably relates to the development of simulation software that use tipping bucket models and simple root water uptake models. Permanent wilting point (PWP) is defined as the medium moisture content at which the leaves of a growing plant first reach a stage of wilting from which they do not recover, it is based on an old study of permanent wilting of dwarf sunflowers [Richard and Weaver, 1943]. Different plants may have different abilities to absorb water at potentials higher than 1500 kPa. However, for many soils or mediums, the changes in water content in this dry range are very small and the water content valuated in 1500 kPa is normally considered the moisture at PWP. The volumetric soil water content varies less in range (around 1500 kPa). The plant-available water (PAW) may be defined as the difference in soil moisture between FC and WP. Studies have shown that biochar amendment can increase the amount of water held at wilting point [Atkinson, 2018; Teixeira *et al.*, 2017].

3.4.1 *Determination of maximum WHC in laboratory soil column studies*

One of the standard methods to determine agronomic WHC is by free drainage (under gravity) for a period of 24h or 48h. This method is sometimes colloquially referred to as "gravity-drained equilibrium water content" [Laird *et al.*, 2010], the "European" Maximum WHC method, field-carrying capacity, drained upper limit and maximum field or capillary capacity. Maximum WHC is generally determined in cylinders packed with soil or amended soil based on the method of Laird *et al.* [2010] and Veihmeyer and Hendrickson [1948].

3.5 Biochar WHC in Different Biomass and Temperatures

A commonly reported effect of biochar application to soil is an improvement in WHC [Blanco-Canqui, 2017; Omondi *et al.*, 2016; Obia *et al.*, 2016; Sun and Lu, 2014; Abel *et al.*, 2013; Liu and Zhang, 2013; Verheijen *et al.*, 2010; Nunes and Teixeira, 2010; Asai *et al.*, 2009; Sohi *et al.*, 2009]. Other studies have corroborated positive effects of improving WHC by adding biochar for different types of soils [Asai *et al.*, 2009; Sun and Lu, 2014] and for biochars produced from different feedstocks and technologies [Laird *et al.*, 2010; Basso *et al.*, 2013]. Novak *et al.* [2012] reported that biochar amendments enhanced the moisture storage capacity of Ultisols and Aridisols, but the effect varied with feedstock selection and pyrolysis temperature. Biochar increased WHC and PAW [Martinsen *et al.*, 2014; Mukherjee and Lal, 2014; Basso *et al.*, 2013; Cornellisen *et al.*, 2013; Herath *et al.*, 2013]. An increase in the PWP after biochar application may cause a reduction in AWC. Despite its generally high porosity, biochar application does not always enhance WHC. Some studies have also shown no effect of biochar addition or reductions in WHC at different potentials [Paneque *et al.*, 2016; Jeffery *et al.*, 2015; Ojeda *et al.*, 2015; Hardie *et al.*, 2014; Major *et al.*, 2012].

Hardie *et al.* [2014] reported no significant effects on soil moisture characteristics or PAW following application of a green waste biochar applied to a clay loam soil. Major *et al.* [2012] also reported no significant effect on WHC of a clay soil following the addition of 20 Mg ha^{-1} of wood biochar. It is remarkable that some studies that reported improved WHC have used biochar application rates that are probably economically unfeasible for large-scale field applications, such as 195 Mg ha^{-1} as reported by Yu *et al.* [2013]. The causal mechanisms of changes in soil water content after addition of biochar are still not completely understood. The most obvious mechanisms that increase soil water retention are related to more water stored within the biochar pore space [Basso *et al.*, 2013], which is generally highly porous. Sun and Lu [2014] speculated that enhanced soil water retention after biochar application was the consequence of an increased aggregate stability. Teixeira *et al.* [2017] showed that increased carbonization temperature of bamboo biomass increases the amount of water held at high potential (>1500 kPa), probably related with an increased specific surface area (SSA). Alterations in zeta potential and enhanced cation exchange capacity were also related with the adsorption of hydrated ions on the biochar surface and correlated with the enhancement of water retention [Batista *et al.*, 2018].

Pore size distribution of biochar may range five orders of magnitude, from sub-nanometer to tens of micrometers [Brewer *et al.*, 2014]. Different methods have been used to quantify and characterize the pores of biochar samples such as mercury intrusion [Brewer *et al.*, 2014] or capillary equation [Liu *et al.*, 2017].

Gray *et al.* [2014] have suggested that the hydrophobicity of biochars may impede uptake of water into the pore space of biochars regardless of pore size and structure. They attributed the hydrophobicity to a negative relationship between the feedstock and biochar production temperature and rate of production of hydrophobic compounds that remain on the surface of the biochar pores. These authors use an interesting approach to estimate biochar hydrophobicity using water and ethanol uptake.

Jeffery *et al.* [2015] found that biochar addition does not improve the hydrological function of a sandy soil, and speculated that biochars that were strongly hydrophobic prevented water from infiltrating inside. Yi *et al*. [2015] found that higher pyrolysis temperatures reduced hydrophobicity of the resulting biochar, from a contact angle of 101° (after 300°C pyrolysis) to 21° (after 600°C pyrolysis). Adding wettable biochars to soils with inherent water repellence has been shown to reduce or even eliminate soil water repellence [Hallin *et al*., 2015; Page-Dumroese, 2015]. Biochar hydrophobicity is a prime candidate for inclusion in existing biochar certification systems [Verheijen *et al*., 2019b].

Biochar particle size effect on WHC was studied by Verheijen *et al*. [2019], Blanco-Canqui *et al*. [2017] and Glab *et al*. [2016]. These studies observed that fine biochar particles increased gravimetric soil water content more efficiently than coarse particles. Liu *et al*. [2017] showed that biochar particle size also affects WHC by changing pore space between particles and by adding pores inside the biochar particles.

The impact of biochar application to soil on PAW will depend on its properties or attributes. Key properties are high porosity and large SSA, which contribute to the adsorptive properties of biochar. In general, SSA increases with an increase in peak temperature of biochar production [Teixeira *et al*., 2017]. At relatively low temperatures of carbonization, the volatiles, tars and other products will remain or fill the internal pore structure of the biomass, and the SSA will be limited [Chen *et al*., 2019]. The effect of carbonization temperatures to enhance SSA reached a limit. At very high temperatures, the SSA is decreased, probably related to clogging of some micropores by volatiles. The fast ramp rate may reduce the effect of pore clogging at high temperatures [Brown *et al*., 2016]. The large SSA may have a positive linear correlation with biochar WHC at high pressures (1500 kPa or higher), but it will not necessarily enhance the PAW.

The use of bamboo feedstock, by using natural occurrence or harvesting plantation, to produce charcoal is an option that is still little studied and discussed. A study Teixeira *et al*. [2017] showed that the increase of carbonization temperature altered its chemical–physical properties and increased the amount of water held with high potentials (>1500 kPa). The bamboo biochar produced with temperatures of 500°C and 550°C showed the higher values of PAW, therefore, using this criterion (high PAW) are the preferable temperatures to produce bamboo biochar for growing media. The ideal temperature of carbonization for production of biochar will depended on the porosity and WHC required for the biochar application. A compilation of studies about WHC of biochar and or soil mixed with biochar is shown in Table 3.2. The feedstocks, pyrolysis technique,

Table 3.2:   Studies of effect of biochar in water retention.

Feedstocks	Carbonization techniques	Biochar rates and textural class of soils	Technique to measure water retention	Range of potentials measured (kPa)	References
Maize (*Zea mais*) (mix of the whole plant)	Slow pyrolysis — 750°C — RT: 20 min; Hydrothermal — 200°C — 20 bar — RT: 10 h; Retort kiln — 550°C — RT: 15 min	0; 10; 25; 50 g kg⁻¹; Sand; loamy sand	Hanging water and pressure plates	0; 6; 10; 32; 1000; 1500	Abel et al. [2013]
Forest chipping trunks (*Abies alba, Larix decidua, Picea excelca, Pinus spp.*)	Slow pyrolysis — 450°C — RT: 48 h	Pure soil and biochar Sand	Büchner funnel and pressure plates	0 – 7.5; 10; 33; 102; 306; 1500	Baiamonte et al. [2019]
Quercus rubra	Fast pyrolysis — 500°C	0; 30; 60 g kg⁻¹ Sandy loam	Pressure plates	1; 2.5; 5; 10; 20; 33; 50; 100; 1500	Basso [2013]
Coconut shells (*Coco nucifera*), orange peels (*Citrus sinensis*), oil palm bunch (*Elaeis guineensis*), sugarcane bagasse (*Saccharum officinarum*), water hyacinth plants (*Eichhornia crassipes*), charcoal fines (*Eucaliptus spp.*)	Tunnel oven — 350°C — RT: 1 h	0; 50 g kg⁻¹ Sand	Gravity drained equilibrium water content	¹WHC	Batista et al. [2018]
Pruning of fruits trees	Slow pyrolysis — 300°C — RT: 15 min	0; 10; 30 g kg⁻¹ Clay	Büchner funnel and pressure plates	0.2; 0.5; 1.0; 2.0; 4.0; 6.0; 8.0;10;12	Castellini et al. [2015]
Straw of *Miscanthus giganteus* and *Triticum aestivum*	Slow pyrolysis 500°C	5; 10; 20; 40 g kg⁻¹ Loamy sand	Pressures plates	4.0; 10; 33; 100; 200; 500; 1500	Głąb et al. [2016]
Picea abies (70%) and *Fagus sylvatica* (30%)	Industrial kiln — 550–600°C	0; 10; 35 g kg⁻¹ Sand	Gravity drained equilibrium water content	Pot water holding capacity	Haider et al. [2014]
Acacia spp.	Flow kiln 600°C RT: 30–40 min	47 Mg ha⁻¹ Sandy loam	Pressure plates and evaporative flux	0; 0.10; 1.0; 3.0; 10; 1500	Hardie et al. [2014]
Maize stover (*Zea mais*)	Rotating kiln — 300–550°C	10; 11.3; 17.3 Mg ha⁻¹ Silty loam	Haines apparatus and pressure plates	2; 4; 6; 8; 10; 30; 100 e 1500	Herath et al. [2013]

North European grassland	Industrial pyrolysis — 400°C and 600°C RT: 5 min	10 Mg ha⁻¹ 1; 5; 20 and 50 Mg ha⁻¹ Sand	Tension tables, pressure plates and pressure cells	0.25; 1; 5; 7; 10; 20; 70; 200; 1500	Jeffery et al. [2015]
Sugarcane bagasse (*Saccharum officinarum*)	Pyrolysis at 400°C, 600°C and 800°C RT: 2 h	30 g kg⁻¹ Clay	Tension table and pressure plate	0 to 30; 30 to 1500	Kameyama et al. [2016]
Wood-chips (cedar; cypress; moso bamboo; rice husk; sugarcane bagasse; poultry manure, agricultural wastewater sludge)	Pyrolysis at 400°C, 600°C and 800°C RT: 2 h	1000 g kg⁻¹ [Pure biochar]	Tension table and pressure plate	0 to 30; 30 to 1500	Kameyama et al. [2019]
Switchgrass (*Panicum virgatum*)	Torrefaction chamber — 375–475°C	10 g kg⁻¹ Silty clay loam; loam; sandy loam; silty clay loam	Pressure plates	0.10; 0.33; 10; 30; 50; 1000; 1500	Koide et al. [2015]
Mixed hardwood of oak (*Quercus* spp.) and hickory (*Carya* spp.)	Slow pyrolysis — traditional kilns	0; 5; 10; or 20 g kg⁻¹ Loam	Gravity drained equilibrium water content a	¹WHC [column]	Laird et al. [2010]
Dairy manure and woodchip	Muffle furnace 300°C, 500°C, 700°C RT: 1 h	50 g kg⁻¹ Loam	Buchner Funnel and SWRC equation	0 to 15; 33; 1500	Lei and Zhang [2013]
Prosopis spp.	Muffle furnace 400°C — RT: 4 hours	0; 20 g kg⁻¹ Sand	Evaporative flux and Psychrometry — Dew Pointcrometry	0 to 44; 100 to 30000	Liu et al. [2017]
Rice husk Biochar (*Oriza sativa*)	Char from thermal industry	0; 20; 40; 60 g kg⁻¹ Clay	Tension table and pressure plate	33; 1500	Lu et al. [2014]
Maize cobs (*Zea mais*)	Earthmound kiln Brick kiln both temperature around 350°C RT: 7 days	0; 2; 6 Mg ha⁻¹ Sandy soils; sandy loamy sand soils; sandy loam, loam	Pressure plates	10; 1500	Martinsen et al. [2014]
Woodship (*Eucalyptus spp.*),	Metal kiln ~ 400–500°C	0; 8; 16; 32 Mg ha⁻¹ Sandy loam; clay	Centrifuge	6.0; 8.0; 10; 33; 60; 100; 1500	Melo Carvalho et al. [2014, 2015]

(*Continued*)

Table 3.2:   (*Continued*)

Feedstocks	Carbonization techniques	Biochar rates and textural class of soils	Technique to measure water retention	Range of potentials measured (kPa)	References
Corn stover (*Zea mais*) and switchgrass (*Panicum virgatum*)	Slow pyrolysis container nested retort — 400–800°C — RT: 4590 min Microwave pyrolysis 650°C — RT: 18 min	40 g kg⁻¹ [104 Mg ha⁻¹] Sandy loam	Tension table and pressure plates Psychrometry — Dew Point	1; 3; 5; 7; 10; 25; 100; 500; 1000; 1500	Mollinedo et al. [2015]
Norway spruce (*Picea albies*) 69%, Beech (*Fagus sylvatica*) 19% plus other wood species	Slow pyrolysis — 480°C	0; 5.4 g kg⁻¹ [0; 20 Mg ha⁻¹] Sandy loam	Tension table and pressure plates	5; 10; 30; 50; 70; 100; 200; 340; 700; 1020; 3000; 15300	Nelissen et al. [2015]
Peanut hulls, pecan shells, poultry litter, switchgrass, hardwood waste	Rotating kiln — 400–500°C RT: 1–2 h Retort — 250–700°C — RT: 1–2 h Fluidized-bed kiln — 500°C — RT: 5s	20 g kg⁻¹ Loamy sand; silt loam	Evaporative flux	5 to 60	Novak et al. [2012]
Tropical wood	Traditional hot tail ~400°C RT: 2–3 days	0, 30, 50, 70% (volume basis) Clay, sand clay loam	Tension tables and pressure plates	0, 10, 32, 60, 100	Nunes and Teixeira [2010]
Maize cob (*Zea mais*) Rice husk (*Oriza sativa*)	Brick kiln — 300–350°C — RT: 2 days Retort — 300–350°C — RT: 1 day	0; 2; 6 Mg ha⁻¹ and 0; 17.5; 35 Mg ha⁻¹ Loamy sand; sand loamy; sand	Tension table and pressure plates	1; 3; 5; 7; 10; 100; 1500	Obia et al. [2016]
Poplar and pine wood splinters, or thermally-dried municipal sewage sludge	Slow pyrolysis — 500–550°C RT: 15 min Fast pyrolysis — 600–900°C — RT: <1 min Gasification — 600–900°C	10 g kg⁻¹ Sandy loam	Tension table and Psychrometry — Dew Point	1; 6; 10; 30; 50; 75; 1500; 30000	Ojeda et al. [2015]

Feedstock	Production conditions	Application rate and soil	Method	¹WHC	Reference
Pine wood, paper-sludge, grapevine wood, sewage sludge	Fast pyrolysis — 500–620°C RT: 20 min Traditional kiln	0; 1.5; 15 Mg ha⁻¹ Sandy loam	Filter paper	¹WHC	Paneque et al. [2016]
Woodmil waste of Corsican Pine (*Pinus nigra*)	Gasification — 1000°C Pyrolysis — 450–500°C RT: 1 h	0.1; 0.5 and 2.5% 0; 10; 50; 60 g kg⁻¹ 4; 20 and 100 Mg ha⁻¹ Loamy sand; silt loam; sandy loam; loam; silty clay loam	Gravity drained equilibrium water content Biological determination of the wilt point	¹WHC [column]	Peake et al. [2014]
Tropical wood	Traditional hot tail ~400°C RT: 2–3 days	500; 1000 g kg⁻¹ Sand and chicken manure	Tension table	0; 1; 3; 6;10; 20	Souza et al. [2002]
Tropical wood	Traditional hot tail ~400°C RT: 2–3 days	30 g kg⁻¹ [3.5 Mg ha⁻¹] Clay	Tension table and pressure plates	0; 1; 6; 10; 100	Steiner et al. [2009]
Straw, woodchips, wastewater-sludge biochar	Factory reactor ~500°C RT: 2 h	0; 20; 40; 60 g kg⁻¹ Clay loam	Pressure plate	33; 1500	Sun and Lu [2014]
Chips of debarked spruce (*Picea abies*) and pine (*Pinus sylvestris*)	Pressurized carbonizer 550–600°C	0; 5; 10 Mg ha⁻¹ Loamy sand	Tension table and pressure plates	0; 3; 6; 10; 50; 250; 1500	Tammeorg et al. [2013, 2014]
Stems of Bamboo (*Phyllostachys pubescens*) and tropical wood	Rotationary kiln for bamboo — 400°C, 450°C, 500°C, 550°C, 600°C and hot tail oven for wood ~400°C	Pure biochar	Tension table and pressure plates	0; 1.0; 3.3; 6.2; 10; 33	Teixeira et al. [2017]
Pine and oak	Clay, sand, clay loam	0; 15; 30; 45% (volume basis)	Equivalent moisture (centrifuge)	Field capacity and Permanent wilting point	Tryon [1948]

(*Continued*)

Table 3.2: *(Continued)*

Feedstocks	Carbonization techniques	Biochar rates and textural class of soils	Technique to measure water retention	Range of potentials measured (kPa)	References
Deciduous mixed wood feedstock of sycamore (*Acer pseudoplatanus*), Oak (*Quercus spp.*), Beech (*Fagus sylvatica*) and Bird Cherry (*Prunus padus*)	Pyrolysis — 600°C RT: 16 h	0; 30; 60 Mg ha^{-1} Sandy loam	Tension table	0; 5; 7.5; 10; 50	Ulyett *et al.* [2014]
Yellow pine	Retort at 400°C — RT: 3 h	0; 10; 20; 33; 40; 50; 60; 70; 80; 90; 100; 150; 200; 250; 300; 350; 400; 450; 500; 750; 1000 g kg^{-1} Loamy sand	Gravity drained equilibrium water content	[1]WHC [pot]	Yu *et al.* [2013]
Mixed wood	Industrial pyrolysis — 620°C RT: 20 min	0; 1; 5; 10; 20% of biochar Sand and sandy loam	Gravity drained equilibrium water content	[1]WHC [column]	Verheijen *et al.* [2019]
Straw, woodchips, wastewater sludge	Industrial slow pyrolysis — 500°C RT: 2 h	0; 20; 40; 60 g kg^{-1} Clay loam	Pressure plates	1; 5; 8; 10; 20; 33; 50; 100; 200; 1000; 1500	Zong *et al.* [2016]

Notes: [1]WHC: Water holding capacity for packed soil in pots or columns. RT: Residential time.
Sources: See details on Laird *et al.* [2014]; Verheijen *et al.* [2019].

textural classes of soil, the potential selected to estimate WHC give an overview of the state of art and guide for some comparisons. We recommend that in future studies about the effect of biochar in water holding capacity, a more detailed description will be given, including details of the feedstock (i.e. plant species and part(s) used for carbonization), carbonization technique (type of kiln, heating ramp, residential time, size of charcoal particles), rate used (i.e. Mg ha^{-1} and also g kg^{-1}), soil texture class, soil classification (i.e. using the local and the IUSS Working Group WRB [2015] system) and at least the water held at saturation (0 kPa), 100, 330 (FC) and 1500 (PWP) kPa. The other parameters recommended for biochar characterization [IBI, 2015; EBC, 2012] are also quite important to allow comparison and to understand possible mechanisms involved.

3.6 Final Remarks

The feasibility of adding biochar to soil to enhance WHC is still controversial not only because of non-significant effects in many soils but also because the feedstock and the pyrolysis technique used to produce biochar and the consequences in WHC are still not completely understood. Moreover, some results indicate that a large amount of biochar is necessary to enhance PAW, reducing economic feasibility for farmers without grants (government or industry) for sequestering or reducing carbon emissions or other ecosystem services. The effect of enhancing soil carbon content (not only from biochar application) to increase PAW in soils is not straightforward [Minany and McBratney, 2018]. The addition of biochar in soil may improve or decrease the FC, PMP and consequently reduce or enhance PAW that has a direct effect on plant growth and/or crop yield. Designed or tailor-made biochar may be produced choosing specific feedstocks and pyrolysis techniques tailored to produce specific characteristics (i.e. explicit pore size distribution and SSA) to maximize soil-water storage in determined ranges of pressures or PAW. Biochar that is designed to enhance secondary soil structure, i.e. aggregation, and thereby WHC and PAW, requires further research. Biochar additions may be more effective to enhance PAW in soil with a coarse texture (i.e. sandy, loamy sandy, sandy loam, sandy clay loam, sandy clay). Caution is necessary in application of large amounts of biochar produced from not well-studied feedstocks or pyrolysis technique that is not produced to comply with the biochar standards or certificates [IBI, 2015; EBC, 2012] as some consequences may not be reversible [Verheijen *et al.*, 2010, 2012, 2019b].

References

Abel, S., Peters, A., Trinks, S., Schonsky, H., Facklam, M. and Wessolek, G. (2013). Impact of biochar and hydrochar addition on water retention and water repellency of sandy soil. *Geoderma*, 202, pp. 183–191.

Abrol, V., Ben-Hur, M., Verheijen, F. G., Keizer, J. J., Martins, M. A., Tenaw, H., Tchehansky, L. and Graber, E. R. (2016). Biochar effects on soil water infiltration and erosion under seal formation conditions: Rainfall simulation experiment. *J. Soils Sediments*, 16(12), pp. 2709–2719.

Ahn, H. K., Richard, T. L. and Glanville, T. D. (2008). Laboratory determination of compost physical parameters for modeling of airflow characteristics. *Waste Management*, 28(3), pp. 660–670.

Andrenelli, M., A. Maienza, L. Genesio, F. Miglietta, S. Pellegrini, F. Vaccari *et al.* (2016). Field application of pelletized biochar: Short term effect on the hydrological properties of a silty clay loam soil. *Agric. Water Manage*, 163, pp. 190–196.

Atkinson, C. J. (2018). How good is the evidence that soil-applied biochar improves water holding capacity? *Soil Use Manag.*, 34(2), pp. 177–186.

Baiamonte, G., Crescimanno, G., Parrino, F. and de Pasquale, C. (2019). Effect of biochar on the physical and structural properties of a desert sandy soil. *Catena*, 175, pp. 294–303.

Baronti, S., Vaccari, F. P., Miglietta, F., Calzolari, C., Lugato, E., Orlandini, S., Pini, R., Zulian, C. and Genesio, L. (2014). Impact of biochar application on plant water relations in Vitis vinifera (L.). *Eur. J. Agronomy*, 53, pp. 38–44.

Basso, A. S., Miguez, F. E., Laird, D. A., Horton, R. and Westgate, M. (2013). Assessing potential of biochar for increasing water-holding capacity of sandy soils. *GCB Bioenergy*, 5, pp. 132–143.

Batista, E. M. C. C., Shultz, J., Matos, T. T. S., Fornari, M. R., Ferreira, T. M., Szpoganicz, B., de Freitas, R. A. and Mangrich, A. S. (2018). Effect of surface and porosity of biochar on WHC aiming indirectly at preservation of the Amazon biome. *Nature Sci. Rep.*, 8, pp. 10677–10686.

Blanco-Canqui, H. (2017). Biochar and soil physical properties. *Soil Sci. Soc. Am. J.*, 81(4), pp. 687–711.

Brewer, C. E., Chuang, V. J., Masiello, C. A., Gonnermann, H., Gao, X., Dugan, B., Driver, L. E., Panzacchi, P., Zygourakis, K. and Davies, C. A. (2014). New approaches to measuring biochar density and porosity. *Biomass Bioenergy*, 66, pp. 176–185.

Brown, R. A., Kercher, A. K., Nguyen, T. H., Nagle, D. C. and Ball, W. P. (2006). Production and characterization of synthetic wood chars for use as surrogates for natural sorbents. *Organic Geochemistry*, 37(3), pp. 321–333.

Castellini, M., Giglio, L., Niedda, M., Palumbo, A. D. and Ventrella, D. (2015). Impact of biochar addition on the physical and hydraulic properties of a clay soil. *Soil Tillage Res.*, 154, pp. 1–13.

Chen, W., Meng, J., Han, X., Lan, Y. and Zhang, W. (2019). Past, present, and future of biochar. *Biochar*, 1(1), pp. 75–87.

Cornelissen, G., Martinsen, V., Shitumbanuma, V., Alling, V., Breedveld, G. D., Rutherford, D. W., Sparrevik, M., Hale, S. E., Obia, A. and Mulder, J. (2013). Biochar effect on maize yield and soil characteristics in five conservation farming sites in Zambia. *Agron J.*, 3, pp. 256–274.

Dane, J. H. and Topp C. G. (2002). Water content. In: Dane, J. H. and Topp, G. C. (eds.), *Methods of Soil Analysis*. Chapter 3.1. SSSA Book Ser. 5. SSSA, Madison, pp. 417–545.

Dane, J. H. and J. W. Hopmans, J. W. (2002). Water retention and storage: Laboratory methods. In: Dane, J. H. and Topp, G. C. (eds.), *Methods of Soil Analysis*. Chapter 3.2.2. SSSA Book Ser. 5. SSSA, Madison, pp. 671–720.

Du, Z., Chen, X., Qi, X., Li, Z., Nan, J. and J. Deng. (2016). The effects of biochar and hoggery biogas slurry on fluvo-aquic soil physical and hydraulic properties: A field study of four consecutive wheat–maize rotations. *J. Soils Sediments*, 16, pp. 2050–2058.

EBC (2012). *European Biochar Certificate: Guidelines for a Sustainable Production of Biochar*. www.european-biochar.org/en.

Głąb, T., Palmowska, J., Zaleski, T. and Gondek, K. (2016). Effect of biochar application on soil hydrological properties and physical quality of sandy soil. *Geoderma*, 281, pp. 11–20.

Gray, M., Johnson, M. G., Dragila, M. I. and Kleber, M. (2014). Water uptake in biochars: The roles of porosity and hydrophobicity. *Biomass Bioenergy*, 61, pp. 196–205.

Grossman, R. B. and Reinsch, T. G. (2002). Bulk density and linear extensibility. In: Dane, J. H. and Topp, G. C. *Methods of Soil Analysis*. Chapter 2. SSSA Book Ser. 5. SSSA, Madison. pp. 201–228.

Haider, G., Koyro, H. W., Azam, F., Steffens, D., Müller, C. and Kammann, C. (2014). Biochar but not humic acid product amendment affected maize yields via improving plant-soil moisture relations. *Plant and Soil*, 395, pp. 141–157.

Hallin, I. L., Douglas, P., Doerr, S. H. and Bryant, R. (2015). The effect of addition of a wettable biochar on soil water repellency. *Eur. J. Soil Sci.*, 66(6), pp. 1063–1073.

Hardie, M., Clothier, B., Bound, S., Oliver, G. and Close, D. (2014). Does biochar influence soil physical properties and soil water availability? *Plant Soil*, 376(1–2), pp. 347–361.

Herath, H. M. S. K., Camps-Arbestain, M. and Hedley, M. (2013). Effect of biochar on soil physical properties in two contrasting soils: An Alfisol and an Andisol. *Geoderma*, 209, pp. 188–197.

International Biochar Initiative (IBI). (2015). Standardized product definition and product testing guidelines for biochar that is used in soil. https://www.biochar-international.org/wp-content/uploads/2018/04/IBI_Biochar_Standards_V2.1_Final.pdf.

IUSS Working Group WRB. (2015). World Reference Base for Soil Resources 2014, update 2015 International soil classification system for naming soils and creating legends for soil maps. *World Soil Resources Reports*, No. 106. FAO, Rome.

Jeffery, S., Meinders, M. B. J., Cathelijne, R. S., Bezemer, T. M., van de Voorde, T. F. J., Mommer, L. and van Groenigen, J. W. (2015). Biochar application does not improve the soil hydrological function of a sandy soil. *Geoderma*, 251, pp. 47–54.

Kameyama, K., Miyamoto, T. and Shiono, T. (2014). Influence of biochar incorporation on TDR based soil water content measurements. *Eur. J. Soil Sci.*, 65(1), pp. 105–112.

Kameyama, K., Miyamoto, T., Iwata, Y. and Shiono, T. (2016). Effects of biochar produced from sugarcane bagasse at different pyrolysis temperatures on water retention of a calcaric dark red soil. *Soil Science*, 181(1), pp. 20–28.

Kameyama, K., Miyamoto, T. and Iwata, Y. (2019). The Preliminary Study of Water-Retention Related Properties of Biochar Produced from Various Feedstock at Different Pyrolysis Temperatures. *Materials*, 12(11), 1732.

Kerré, B., Willaert, B., Cornelis, Y. and Smolders, E. (2017). Long-term presence of charcoal increases maize yield in Belgium due to increased soil water availability. *Eur. J. of Agronomy*, 91, pp. 10–15.

Kibblewhite, M. G., Ritz, K. and Swift, M. J. (2007). Soil health in agricultural systems. *Philosophical Transactions of the Royal Society B: Biological Sciences*, 363(1492), pp. 685–701.

Koide, R. T., Nguyen, B. T., Skinner, R. H., Dell, C. J., Peoples, M. S., Adler, P. R. and Drohan, P. J. (2015). Biochar amendment of soil improves resilience to climate change. *Gcb Bioenergy*, 7(5), pp. 1084–1091.

Laird, D., Fleming, P., Wang, B., Horton, R. and Karlen, D. (2010). Biochar impact on nutrient leaching from a Midwestern agricultural soil. *Geoderma*, 158, pp. 436–442.

Lei, O. and Zhang, R. (2013). Effects of biochars derived from different feedstocks and pyrolysis temperatures on soil physical and hydraulic properties. *J. Soils Sediments*, 13(9), 1561–1572.

Liu, Z., Dugan, B., Masiello, C. A. and Gonnermann, H. M. (2017). Biochar particle size, shape, and porosity act together to influence soil water properties. *PLoS One*, 12(6), e0179079.

Lu, S. G., Sun, F. F. and Zong, Y. T. (2014). Effect of rice husk biochar and coal fly ash on some physical properties of expansive clayey soil (Vertisol). *Catena*, 114, pp. 37–44.

Ma, N., L. Zhang, Y. Zhang, L. Yang, C. Yu, G. Yin, *et al.* (2016). Biochar improves soil aggregate stability and water availability in a mollisol after three years of field application. *PLoS One*, 11(5), e0154091.

Major, J., Rondon, M., Molina, D., Riha, S. J. and Lehmann, J. (2012). Nutrient leaching in a Colombian savanna Oxisol amended with biochar. *J. Environ. Qual.*, 41(4), pp. 1076–1086.

Martinsen, V., Mulder, J., Shitumbanuma, V., Sparrevik, M., Børresen, T. and Cornelissen, G. (2014). Farmer-led maize biochar trials: Effect on crop yield and soil nutrients under conservation farming. *J. Plant Nutr. Soil Sc.*, 177(5), pp. 681–695.

Melo Carvalho, M. T. de, Holanda Nunes Maia, A. de, Madari, B. E., Bastiaans, L., Van Oort, P. A. J., Heinemann, A. B. and Meinke, H. (2014). Biochar increases plant-available water in a sandy loam soil under an aerobic rice crop system. *Solid Earth*, 5(2), pp. 939–952.

Melo Carvalho, M. T. de, Madari, B. E., Bastiaans, L., Van Oort, P. A. J., Leal, W. G. O., Heinemann, A. B. and Meinke, H. (2016). Properties of a clay soil from 1.5 to 3.5 years after biochar application and the impact on rice yield. *Geoderma*, 276, pp. 7–18.

Minasny, B. and McBratney, A. B. (2018). Limited effect of organic matter on soil available water capacity. *Eur. J. Soil Sci.*, 69(1), pp. 39–47.

Mukherjee, A. and Lal, R. (2014). The biochar dilemma. *Soil Res.*, 50, pp. 217–230.

Nelissen, V., Ruysschaert, G., Manka'Abusi, D., D'Hose, T., De Beuf, K., Al-Barri, B., Cornelis, W. and Boeckx, P. (2015). Impact of a woody biochar on properties of a sandy loam soil and spring barley during a two-year field experiment. *Eur. J. Agronomy*, 62, pp. 65–78.

Nunes, M. M. and Teixeira, W. G. (2010). Growing of Brasil Nuts seedling (*Bertholletia excelsa H.B.K.*) in response to biochar levels as growing media [Crescimento de mudas de castanheiras do Brasil (*Bertholletia excelsa H.B.K.*) em função de doses de carvão vegetal como componente de substrato (Spanish)]. In: III Reunião Científica da Rede CTPetro Amazônia, 2010, Manaus. Anais da III Reunião Científica da Rede CTPetro Amazônia. Manaus: INPA, 2010. v. CD ROM.

Novak, J. M., Busscher, W. J., Watts, D. W., Amonette, J. E., Ippolito, J. A., Lima, I. M., Gaskin, J., Das, K. C., Steiner, C., Ahmedna, M., Rehrah, D. and Schomberg, H. (2012). Biochars impact on soil-moisture storage in an Ultisol and two aridisols. *Soil Sci.*, 177, pp. 310–320.

Obia, A., Mulder, J., Martinsen, V., Cornelissen, G. and Børresen, T. (2016). In situ effects of biochar on aggregation, water retention and porosity in light-textured tropical soils. *Soil and Tillage Res.*, 155, 35–44.

Ojeda, G., Mattana, S., Ávila, A., Alcañiz, J. M., Volkman, M. and Bachmann, J., (2015). Are soilwater functions affected by biochar application? *Geoderma*, 249–250, pp. 1–11.

Omondi, M. O., Xia, X., Nahayo, A., Liu, X., Korai, P. K. and Pan, G. (2016). Quantification of biochar effects on soil hydrological properties using meta-analysis of literature data. *Geoderma*, 274, pp. 28–34.

Page-Dumroese, D. S., Robichaud, P. R., Brown, R. E. and Tirocke, J. M. (2015). Water repellency of two forest soils after biochar addition. *Transactions of the ASABE*, 58(2), pp. 335–342.

Paneque, M., José, M., Franco-Navarro, J. D., Colmenero-Flores, J. M. and Knicker, H. (2016). Effect of biochar amendment on morphology, productivity and water relations of sunflower plants under non-irrigation conditions. *Catena*, 147, pp. 280–287.

Peake, L. R., Reid, B. J. and Tang, X. (2014). Quantifying the influence of biochar on the physical and hydrological properties of dissimilar soils. *Geoderma*, 235–236, pp. 182–190.

Romano, N. and Santini, A. (2002). Water retention and storage: Field. In: Dane, J. H. and Topp, G. C. (eds.), *Methods of Soil Analysis*. Chapter 4. SSSA Book Ser. 5. SSSA, Madison, pp. 721–738.

Richards, L. A. and Weaver, L. R. (1943). Fifteen-atmosphere percentage as related to the permanent wilting percentage. *Soil Sci.*, 56(5), pp. 331–340.

Sohi, S., Lopez-Capel, E., Krull, E. and Bol, R. (2009). Biochar, Climate Change and Soil: A Review to Guide Future Research. CSIRO Land and Water Science Report 05/09.

Steiner, C., Teixeira, W. G. and Zech, W. (2009). Effects of charcoal in Banana (*Musa sp*) planting holes — an on-farm study in central Amazonia, Brazil. In: Woods, W. I., Teixeira, W. G., Lehmann, J., Steiner, C., WinklerPrins, A. and Rebellato, L. (eds.). Amazonian Dark Earth: Wim Sombroek's Vision. (Springer), pp. 423–432.

Sun, F. and Lu, S. (2014). Biochars improve aggregate stability, water retention, and pore-space properties of clayey soil. *Z. Pflanzenernähr. Bodenkd*, 177(1), pp. 26–33.

Tammeorg, P., Simojoki, A., Mäkelä, P., Stoddard, F. L., Alakukku, L. and Helenius, J. (2013). Biochar application to a fertile sandy clay loam in boreal conditions: Effects on soil properties and yield formation of wheat, turnip rape and faba bean. *Plant Soil*, 374, pp. 89–107.

Tammeorg, P., Simojoki, A., Mäkelä, P., Stoddard, F. L., Alakukku, L. and Helenius, J. (2014). Short-term effects of biochar on soil properties and wheat yield formation with meat bone meal and inorganic fertiliser on a boreal loamy sand. *Agric. Ecosyst. Environ*, 191, pp. 108–116.

Tammeorg, P., Bastos, A. C., Jeffery, S., Rees, F., Kern, J., Graber, E. R., Ventura, M., Amaro, A., Budai, A., Cordovil, C. M. D. S., Domene, X., Gardi, C., Gascó, G., Horák, J., Kammann, C. I., Kondrlova, E., Laird, D., Loureiro, S., Martins, M. A. S., Panzacchi, P., Prasad, M., Prodana, M., Puga, A. P., Ruysschaert, G., Saspaszt, L., Silva, F. C., Teixeira, W. G., Tonon, G., Delle Vedove, G., Zavalloni, C., Glaser, B. and Verheijen, F. G. A. (2017). Biochars in soils: Towards the required level of scientific understanding. *J. Environ. Eng. Landsc*, 25(2), pp. 192–207.

Teixeira, W. G., Marques, J. D. O., Steiner, C. and Flanagam, R. (2017). Water retention in bamboo and wood biochar produced at different temperatures. In: Drumond, P. M. and Wiedman, G. (eds.), *Bambus no Brasil: da biologia à tecnologia*. Rio de Janeiro: Instituto Ciência Hoje — Embrapa, pp. 368–380.

Teixeira, W. G. and Behring, S. B. (2017). Retenção de água no solo pelos métodos da mesa de tensão e da câmara de Richards. In: Teixeira, P. C., Donagemma, G. K., *et al*. (eds.). Manual de métodos de análise de solo. 3. (Embrapa, Brasília), pp. 34–46.

Tryon, E. H. (1948). Effect of charcoal on certain physical, chemical, and biological properties of forest soils. *Ecological Monographs*, 18(1), pp. 81–115.

Ulyett, J., Sakrabani, R., Kibblewhite, M. and Hann, M. (2014). Impact of biochar addition on water retention, nitrification and carbon dioxide evolution from two sandy loam soils. *Eur. J. Soil Sc.*, 65(1), pp. 96–104.

Verheijen, F., Jeffery, S., Bastos, A. C., Van der Velde, M. and Diafas, I. (2010). Biochar application to soils. *A Critical Scientific Review of Effects on Soil Properties, Processes, and Functions. EUR*, 24099, p. 162.

Verheijen, F. G., Montanarella, L. and Bastos, A. C. (2012). Sustainability, certification, and regulation of biochar. *Pesq. Agrop. Bras.*, 47(5), pp. 649–653.

Verheijen, F. G., Zhuravel, A., Silva, F. C., Amaro, A., Ben-Hur, M. and Keizer, J. J. (2019a). The influence of biochar particle size and concentration on bulk density and maximum water holding capacity of sandy vs sandy loam soil in a column experiment. *Geoderma*, 347, pp. 194–202.

Verheijen, F. G., Bastos, A. C., Schmidt, H. P. and Jeffery, S. (2019b). *Biochar and certification. Sustainability Certification Schemes in the Agricultural and Natural Resource Sectors: Outcomes for Society and the Environment*, p. 113.

Veihmeyer, F. J. and Hendrickson, A. H. (1948). The permanent wilting percentage as a reference for the measurement of soil moisture. *Eos, Transactions American Geophysical Union*, 29(6), pp. 887–896.

Wang, T., Stewart, C. E., Ma, J., Zheng, J. and Zhang, X. (2017). Applicability of five models to simulate water infiltration into soil with added biochar, *J. of Arid Land*, 9(5), pp. 701–711.

Yi, S., Witt, B., Chiu, P., Guo, M. and Imhoff, P. (2015). The origin and reversible nature of poultry litter biochar hydrophobicity. *J. of Environ. Quality*, 44(3), pp. 963–971.

Young, M. H. and Sisson, J. B. (2002). Tensiometry. In: Dane, J. H. and Topp, G. C. (eds.), *Methods of Soil Analysis*: Part 4 physical methods. SSSA Book Ser. 5. Madison, pp. 575–678.

Yu, O.Y., Raichle, B. and Sink, S. (2013). Impact of biochar on the water holding capacity of loamy sand soil. *Int. J. Energy Environ. Eng.*, 4(44), pp. 1–9.

Zong, Y., Xiao, Q. and Lu, S. (2016). Acidity, water retention, and mechanical physical quality of a strongly acidic Ultisol amended with biochars derived from different feedstocks. *J. Soils Sediments*, 16(1), pp. 177–190.

Chapter 4

Biochar as a Growing Media Component

Agnieszka Medyńska-Juraszek* and Irmina Ćwieląg-Piasecka

Institute of Soil Science and Environmental Protection, Wrocław University of Environmental and Life Sciences, Poland

*agnieszka.medynska-juraszek@upwr.edu.pl

Abstract

The chapter describes biochar use as growing media in horticulture. Soilless production has become increasingly important, bringing new challenges to meet with expectations of consumers and policies. Biochar as an organic waste-based highly porous material, with great sorption capacity for nutrient and water storage gives "win–win" solution for horticulture in terms of adjustment to climate change and reduction of environmental impacts, offering a number of benefits to growers. Reduced water and mineral fertilizer use, plant growth stimulation or higher and better quality yields are the most promising benefits. The chapter summarizes the state of knowledge about biochar use in soilless production of horticultural crops, based on scientific reports and results from the experiments conducted worldwide. The provided knowledge can be helpful to answer frequently asked questions: *Is it possible to replace standard growing medium with biochar* and *what are the benefits of peat or mineral wool replacement with biochar for my crop production?*

4.1 Introduction

Soilless production has become increasingly important globally over the last 60 years. Compared with soil-based cultivation, soilless or hydroponic production can be more cost-effective, due to higher water and nutrient use efficiencies. Containerization of plant production is well-known to give a profit of higher yields from smaller area. However, soilless production also has some challenges. First, a container environment provides very limited space for root system development, and second, limited water and nutrient storage capacity. Essentially, an effective growing medium must provide plants with nutrients, air and water, and physical support for root development. Plant growing media must have well-balanced physical and chemical properties such as pH, electrical conductivity (EC), C:N mass ratio, bulk density (BD), air filled porosity (AFP), water holding capacity (WHC) and stability, sustaining a favorable balance between air and water storage. In addition, the growing media must be free of pathogens and toxic chemicals, providing good conditions for beneficial living organism development. A key factor contributing to the increased yields obtained from soilless production, as opposed to soil-based production, is the opportunity to select substrate materials that optimize and support plant growth. There is a constant search for materials that have optimal properties for plant growth, are easier to handle, obtain and originate from readily recyclable waste materials. There is a growing interest in the use of suitable, locally sourced materials to minimize the environmental impact and economic costs associated with transport.

4.2 Biochar as a Substitute for Peat

Many different substrates such as sphagnum peat, coir peat, bark, compost, charcoal, perlite and rockwool have been commonly used by the growers to formulate proper growing media. Among the numerous organic materials used as substrates for soilless cultivation of horticultural crops, peat is currently a major component of containerized mixtures for commercial plant production. Low content of nutrients, but on the other hand, its ability to adsorb and release them when added as fertilizer, made peat an ideal constituent of soilless growing media. Widespread reserves of peat make it a readily available and relatively cheap resource. In the last decade, there has been an intensifying pressure from society on growers to reduce the environmental impact of plant production. Becoming more environmentally aware, using peat as a growing media became a global concern. Extraction of peat has many negative impacts on the environment, most important of these are climate change exacerbation and non-renewable resources impoverishment. In the United Kingdom, consumption of peat leads to the annual carbon dioxide emissions of more than 630,000 tonnes, at a cost to society of £32.5 million, arising from the climate change impact. Concerns about the impact of peat extraction on important wildlife habitats and the climate implications of removing the ancient natural carbon stores, have led to calls for a halt to peat extraction and come under increasing scrutiny throughout Europe and particularly in the UK.

Nowadays, the search for alternative high-quality and low-cost materials as growing media in horticulture is a necessity due to the increasing demand and rising costs. In this context, there has been an emphasis on organic materials derived from agricultural, industrial and municipal waste streams [Barrett *et al.*, 2016]. The disposal of waste materials brings many environmental problems, and their re-use or recycle as a growing substrate can be considered as a good solution. Biochar is a carbon-rich coproduct resulting from pyrolyzing of waste biomass. Many research papers have highlighted the benefits of adding biochar to agricultural soils [Glaser *et al.*, 2000; Chan *et al.*, 2007; Lehmann *et al.*, 2011; Warnock *et al.*, 2010; Cárdenas-Aguiar *et al.*, 2017]. These benefits include the promotion of plant growth, the improvement of soil water-holding capacity, diminishing disease incidence in crops, limiting the bioavailability of heavy metals [Park *et al.*, 2010], reducing soil N_2O emission and reducing of nutrient leaching loss, which in turn can reduce fertilizer needs [Laird *et al.*, 2010]. Because biochar is a coproduct of bioenergy production and can contribute to carbon sequestration goals, while also simultaneously increasing yield and reducing fertilizer use, biochar has been touted as a "win–win–win" solution to meeting global environmental challenges.

In the last decades, biochar has attracted widespread attention as a growing media component. Santiago and Santiago [1989] briefly summarized their work using wood-based charcoal chips for hydroponic culture in humid tropical regions of Asia, but they provided few details other than that plants grew well when fertilized with resin-coated fertilizers. Kadota and Niimi [2004] reported 10% or 30% additions of biochar combined with either pyroligneous acid (wood vinegar) or barnyard manure to a 2:1:1:1:1 peatmoss:soil:vermiculite:perlite:sand (v/v) substrate had either no effect or minor changes (positive and negative) in growth parameters of several bedding plant species. Graber *et al.* [2010] reported that biochar improved growth and productivity of pepper (*Capsicum annuum* L.) and tomato (*Lycopersicum esculentum* Mill.) plants in a blend of coconut fiber and tuff, and attributed improvements to either stimulated shifts in microbial populations toward beneficial plant growth-promoting rhizobacteria or fungi or low doses of phytotoxic biochar chemicals, which may have stimulated plant growth at low doses [Altland and Locke, 2013]. Ruamrungsri *et al.* [2011] reported that gloriosa lily (*Gloriosa rothschildiana* L.) in a 1:1:1 sand : rice husk charcoal : coconut fiber substrate did not respond to varying levels of applied calcium (Ca) fertilizers as a result of high Ca levels in rice husk charcoal. Dumroese *et al.* [2011] evaluated pelletized biochar (pellets were 43% biochar, 43% wood flour, 7% polyacetic acid and 7% starch) in combination with sphagnum peatmoss for production of forest seedlings. They found that amendment with 25% biochar pellets improved hydraulic conductivity and water retention at high matric potentials and beneficially increased substrate pH, although concern was noted about lower cation exchange capacity (CEC) and higher carbon : nitrogen ratio. Zhang *et al.* [2014] concluded that the highest quality growth medium and the highest quality ornamental plant growth were achieved by mixing composted green waste with 30% of biochar and 0.7% of humic acids. Vaughn *et al.* [2015] studied the use of biochar as replacements for inorganic components such as vermiculite, perlite and digestate to replace organic components such as peat. They deduced that biochar can substitute peat at levels lower than 15% (v/v). Nieto *et al.* [2016] found that addition of pruning waste biochar in high ratios such as 50 and 75 vol.% to brown peat improves some of their properties as growing media. Margenot *et al.* [2018], by evaluating an alkaline BC at high substitution rates in soil-free substrates with a common ornamental plant, confirmed the feasibility of replacing Sphagnum peat moss with BC for greenhouse and nursery plant production. Full substitution of BC (soft-wood, pH 10.9) for peat in soil-free substrate did not have negative impacts on marigold growth and flowering. Plant responses to the addition of biochar to substrates can be similar to those found for standard substrates containing peat [Dumroese *et al.*, 2011; Altland and Locke, 2012; Vaughn *et al.*, 2013] with profits of creating a beneficial environment for microbes [Graber *et al.*, 2010; Conversa *et al.*, 2015].

4.3 Changing Waste Biomass to Biochar

There are many different and easy ways to make biochar from a wide range of feedstock including green waste (pruning wastes, wood shaving, bark, grass), harvest leftovers (straw, husks, cobs, grain dust, nut shells), vegetable production (old shoots, fruits, peels), leftovers from food production (pulps and residues), paper mill wastes, cotton and wood fibers, sewage sludge and manure. Source material and pyrolysis conditions introduce significant variation in the structure, nutrient content, pH and phenolic content of the biochar products [Novak *et al.*, 2009]. Feedstock selection and pyrolysis conditions affect biochar properties. By understanding and controlling these factors, it is possible to create value-added "designer biochars" for specific applications.

Generally, plant-derived biochars like those obtained from woody biomass or grasses are low nutrient materials due to their scarce ash, nitrogen, phosphorous and potassium contents and have no fertilizing effect. Animal or sewage sludge-derived biochars are more nutrient-reach and can offer additional benefits not only as a conditioner but also as a fertilizer. Biochars produced at lower temperatures generally have lower pH values, ash contents, porosity and specific surface area (SSA) [Lehmann *et al.*, 2011; Spokas *et al.*, 2012; Buecker *et al.*, 2016] than those prepared at higher temperatures, and likewise, higher cation exchange capacities (CECs) and volatile matter contents [Spokas *et al.*, 2011].

Conversion of organic substrates to biochar, which is a recalcitrant (stable) material, is a promising solution for inert and consistent organic origin growing substrates. Biochar production is sustainable and brings a good solution for agricultural animal and crop wastes management, offering a number of benefits to growers. Use of greenwaste biochar in glasshouse production can create a closed loop system whereby biochar produced from crop waste is used as a substrate for soilless production. Turning only tomato crop green waste to biochar would divert 10–60 t·ha^{-1} of green waste away from land fill each crop cycle and would meet 13–50% of the grower's substrate requirement, on per hectare basis [Dunlop *et al.*, 2015]. Growers can also benefit from supplementing glasshouse energy requirements by syngas and bio-oil production during biomass pyrolysis.

Use of biochar in horticulture, especially hydroponics is limited. Many toxic compounds (e.g. polycyclic aromatic hydrocarbons (PAHs), volatile organic compounds (VOCs), heavy metals) are likely to be formed during biochar production. These compounds may exist in biochar matrix and even be bioavailable to exposed plants [Gascó *et al.*, 2016]. Therefore, the formation and impact factors as well as the concentrations and bioavailability of these compounds are reviewed. There is an urgent need to standardized the properties of biochar, which can be selected for growing media components, especially in case of consumption plants. Content of toxic compounds in the material depends mostly on the feedstock type and pyrolysis parameters. The conclusion in case of horticulture is to use biochars produced only from plant-derived materials which are safer and the content of toxic compounds is usually lower compared to sewage sludge, biodegradable municipal wastes or manure biochars. Biochar producer needs to be aware that the product dedicated for horticulture application has to be of premium quality and that there is a necessity to know the source of the biomass and control the content of toxic compounds, e.g. heavy metals, PAHs and residues of pesticides in feedstock before biochar production.

4.4 Water Retention

Producers of containerized plants face several challenges related to water use and runoff. Irrigation must be applied more frequently in containerized production systems compared to field soils, because plant available water is lower within containers filled with soilless substrates, which have high porosity and restricted root volumes. Any water or agrochemicals (i.e. fertilizers, pesticides, and plant growth regulators) applied in an excess of the capacity of the container that are unable to be utilized by the plant or fall outside of the container will likely leach and run off and may eventually impact surface water and the groundwater. Water shortage is a great threat to crop production sustainability and food security. Water retention in biochar also depends strongly on the particle size of the material. Blok *et al.* [2016] found that WHC of the biochar (produced from wood chips) ranges from 56% to 25% v/v, while peat had a WHC ranging from 74% to 35% v/v at suction forces of -10 and -50 cm, respectively. However, once the biochar is mixed with peat, such pores are filled and the WHC is similar to the WHC of the peat alone. Under nearly saturated conditions, the peat and peat-biochar mixture contain at least 20% v/v air, meaning that the rooting media have a sufficient amount of oxygen. Biochars can be considered porous media. Typically, biochar porosity has been classified following IUPAC conventions by distinguishing between micropores (<2 nm), mesopores (2–50 nm) and macropores (>50 nm). Water uptake in porous media depends on capillary forces, which can act to enable or prevent water entry into pores. The strength of capillary forces depends on the surface chemistry and physical properties of the media. Total water uptake of biochar media depends both on capillary forces as well as total porosity. Biochar particle size affects water and nutrient retention potential. Smaller sized particles have a greater surface area that may increase nutrient retention and WHC [Mollinedo *et al.*, 2015]. Depending on the feedstock selection and production conditions, biochars exhibit large ranges in porosity and surface chemistry. Water uptake by biochars is dependent on both feedstock selection, which controls residual macroporosity, and production temperature, which controls hydrophobicity and pyrogenic nanopore formation. With increasing production temperature, residual macroporosity remains relatively constant while pyrogenic nanoporosity increases dramatically. Pyrogenic nanopores contribute only minimally to the total porosity, even in higher temperature biochars, but these pores provide the majority of biochar surface area critical for contaminant and nutrient sorption. Saturation of pores depends on the surface chemistry, which can render biochar surfaces hydrophobic, creating negative capillary pressure that prevents water from entering pores. Overall, hydrophobicity in biochars decrease with increasing production temperature and may be due to aliphatic functionality which is volatilized and lost at higher production temperatures. Reducing particle size and increasing production hold times may also help to reduce hydrophobicity, by allowing greater loss of hydrophobic compounds from biochar particles during pyrolysis. Hydrophobicity is most relevant in external pores and residual macropores, as pyrogenic nanopores are likely filled by water vapor sorption when biochars are submerged in water [Gray *et al.*, 2014]. The hydrophobicity of biochar can be engineered by appropriate choices of feedstock and pyrolysis conditions. Biochar hydrophobicity and field capacity are controllable properties of biochars, with both reaching optimal values (i.e. high field capacity and low hydrophobicity) at pyrolysis temperatures between 400°C and 600°C [Kinney *et al.*, 2012]. WHC of biochar-based growing media can be also easily modified by mixing with relatively wet rooting materials like peat or coir as well as the use of finer grade biochars, e.g. wood-derived.

4.5 Effect on Plant Growth and Crop Quality

Biochar has been shown to promote plant productivity and yield though several mechanisms, but specific mechanisms underlying the contribution of biochar to plant response are poorly understood. Biochar can improve plant productivity directly as a result of its nutrient content and release characteristics, as well as indirectly via improvement of pH, retention of nutrients and water, increased exchange capacity, effects on P and N transformation and turnover, promotion of mycorrhizal fungi and alternation of soil microbial populations and functions [Elad *et al.*, 2010; Graber *et al.*, 2010; Laird *et al.*, 2010; Warnock *et al.*, 2010; Anderson *et al.*, 2011; Lehmann *et al.*, 2011; Aller *et al.*, 2017].

The plant nutrient values of biochar depend on how much biochar is added to growing medium, the properties of the nutrient within the biochar, and the interaction of the biochar and plant. Nutrient contents in biochars are determined greatly by the feedstock source and pyrolytic temperature [Hossain *et al.*, 2011]. Plant-based biochars are nutrient poor and cannot serve as a long-term source of nutrients for container plants. The contents of available N (water-soluble) in biochars decrease with the increase of pyrolysis temperatures from $350°C$ to $600°C$, which could be attributed to the loss of total N and the heterocyclization of N during pyrolysis [Zhang *et al.*, 2016].

Even if nutrients are present in biochar, their availability to plants depends on the type of nutrient and type of biochar. For example, phosphorus is found mainly in the mineral fraction of biochar and only a small share is likely to be associated within the organic structure of biochar. Its solubilization is pH-dependent. Biochar addition to growing media can increase available P, which is associated to the positively charged sites in biochars that increase soil/media capacity to retain and exchange phosphate ions [Jiang *et al.*, 2015]. Phosphorus, in the form of phosphate, is becoming the most expensive fraction of complete fertilizers as its availability becomes more limited. With increasing demand for P and K fertilizers and rising costs of their use, chemical properties suggest that biochar has potential to replace P and K fertilizers dissolved in the irrigation stream [Prasad *et al.*, 2017]. Potassium in biochar is generally readily available to plants. Increased K levels due to biochar application is of great importance as it affects positive fruit quality parameters, such as soluble solids in watermelon or cucumbers, while tomato plants are considered as high K demanding crops.

Locke *et al.* [2013] found that gasified rice hull biochar affects nutrition of horticultural crops in container substrates and contains sufficient P and K to serve as the sole source of nutrients in the short-term (up to 6 weeks) production of several greenhouse crops. Foliar P concentration increased linearly with increasing rice husk biochar amendment (5% and 10%). Altland and Locke [2012] in the column study found that gasified rice husk biochars have sufficient K to supply all the sunflower and geranium. Nitrogen availability from biochars has been shown to vary widely depending on the final temperature of pyrolysis, heating rate, time of holding at the final temperature and type of feedstock. Nitrogen can occur in many forms within biochar, including heterocyclic nitrogen (N contained within the carbon structure of biochar), an organic form of nitrogen which would only become plant available as it is mineralized by microorganisms. In general, availability of nitrogen from biochar is very low. Biochar is usually reach in Ca, Mg and Si, which are mostly in forms available for plants. Micronutrients like B, Cl, Cu, Fe, Mn, Mo and Zn are also present in biochars depending on feedstock type. The actual availability of these nutrients to plants may be associated with the amounts of the total element content. During pyrolysis, residual volatile matter and a big variety of different organic compounds are also formed. These include nitrogen-heterocyclics, substituted furans, phenols and substituted phenols, benzene and substituted benzene, carboxylics and aliphatics. Organic molecules contained in biochar may also have agronomic value as plant growth stimulators. Güereña *et al.* [2015] suggested that organic molecules can enhance plant germination and, what is more, some of them are similar in structure to well-known microbial signaling compounds which have stimulatory effect on nodule formation and N fixation. Biochar addition to a growing substrate may also induce ethylene production by microbes, and as a plant hormone may stimulate or regulate the ripening of fruit, the opening of flowers and the abscission (or shedding) of leaves [Spokas *et al.*, 2012].

Biochar can significantly affect the main chemical and physical characteristics of the potting mixtures that are crucial for nutrient and water uptake effectiveness in plants and prevention of nutrient losses in the system. With increasing amount of biochar added to a growing media pH, EC and BD will increase due to BC high alkalinity and salinity (Ca, Mg, K and Si hydroxides and carbonates present in the biomass). Biochar affects nutrient cycles and their availability mainly by pH modification or direct sorption/desorption processes on biochar surface. Biochar is commonly alkaline and addition to acidic peat can ultimately shift the media status closer to the optimal pH range for the nutrient uptake by plants without need of typical liming procedures. Inorganic carbonates are the major alkaline components of the biochars generated at high temperature [Fidel *et al.*, 2017]. Organic anions also contribute to alkalinity of the biochars, especially for those generated at low temperature. pH and salinity of biochar are easy-to-adjust properties of the material. High pH values of biochar (pH from 8 to 11) can be easily reduced by water rinsing or fertigation with acidic fertilizers.

Biochar amendment can significantly affect plant growth, modifying physical properties of growing medium. Porous structure and better water retention conditions boost root and above vegetative organ development. Graber *et al.* [2010] observed enhanced growth of pepper (*Capsicum annuum* L.) and tomato (*Lycopersicum esculentum* Mill.) in the biochar treated pots. Tomato plant heights were even 39% taller when biochar was added in 3% v/v, compared with the control treatment without biochar. In the same experiment, greater number of nodes was observed at pepper plants. Biochar addition also effects leaf area and length, buds, fruit yields and total fruit weight. Méndez *et al.* [2017] found that lettuce grew better in a peat–biochar mixture than in 100% peat. Nieto *et al.* [2016] observed that lettuce biomass was substantially higher with peat mixed with biochar samples obtained from pyrolysis of pruning waste at 300°C and especially 500°C, than with peat alone.

Biochar amendment can significantly affect leaf nutrient concentration, which was found in many studies on different plant species e.g. pelargonium [Conversa *et al.*, 2015], marigolds [Vaughn *et al.*, 2018], tomato, pepper and cucumber [Graber *et al.*, 2010]. Biochar addition to peat or perlite has positive effect on root growth and development. In perlite–biochar mixtures, with increasing amount of biochar added to the media, fresh mass of cucumber and tomato roots increased by 150% and 80%, respectively. Yields of tomato and cucumber were similar or higher compared to the control on 100% perlite or rockwool. Higher yields of cucumber with higher vitamin C and mineral ash content suggest that biochar can be a good constituent of growing medium for plants with high water demands like watermelon, cantaloupe or leafy vegetables. Biochar addition can also affect above-ground biomass and development of vegetative and reproductive plant parts. In the same experiment, even 50% higher biomass of leafs and length of cucumber stems was observed when 50% of biochar was added to the growing medium. In biochar treated trials, acceleration of early fruit development was observed on tomato. Increase of vegetative part yield may have positive effect on production of leafy vegetables like lettuce, dill, spinach or cabbage [Awad *et al.*, 2017].

Given the complexity of these interactions, it is difficult to isolate those factors that actually play an instrumental role in "biochar effect". Many of these effects are interrelated and may act synergistically to improve crop performance.

4.6 Role of Biochar in Disease and Pest Control

Biochars may act in a similar way in suppressing plant diseases as it was described for other organic soil amendments, such as composts [Lehmann *et al.*, 2011]. Biochars can significantly reduce disease severity of different fungal foliar pathogens with differing infection strategies, by mediating plant systemic resistances in host plants. Some of the main defense-enhancing mechanisms that are currently discussed [Elad *et al.*, 2010; Harel *et al.*, 2012; Schmidt *et al.*, 2014; Graber *et al.*, 2015] are: (a) improved nutrient supply of the host plant, (b) alterations of microbial biomass, as well as community structure and/or function, (c) adsorption and inactivation of toxins produced by the pathogen or other infection-relevant substances, (d) input of biochar-inherent substances that either directly affect the pathogenic organism (e.g. toxins) or induce system-wide defense responses in the host plant. In addition, biochar soil applications are accompanied by several physicochemical alterations of the soil medium (pH, WHC, CEC, BD, soil aeration) which, in turn, affect the above-mentioned mechanisms [George *et al.*, 2016].

Various biochars has been reported to reduce diseases caused by foliar fungal pathogens in tomato, sweet pepper and strawberry [Elad *et al.*, 2010; Harel *et al.*, 2012] and to reduce diseases caused by soilborne pathogens in asparagus, tomato, maple and oak [Elmer *et al.*, 2011; Zwart and Kim, 2012]. Graber *et al.* [2010] observed that the rhizosphere of biochar-amended pepper plants had significantly greater abundance of culturable microbes belonging to the prominent soil-associated groups. In addition, plants grown in biochar amended treatments were shown to be less susceptible to two foliar fungal pathogens (*Bortrytis cinerea* and *Leveillula taurica*), and for pepper, to foliar mite (*Polyphagotarsonemus latus*). De Tender *et al.* [2016] observed that addition of 3% w/w oak wood pyrolysis biochar to peat resulted in a higher fresh and dry plant weight of strawberry, lower susceptibility for the fungal pathogen *Botrytis cinerea* and changes in the rhizosphere microbiology, such as an increase of bacterial diversity and a shift in composition of the rhizosphere micro-biota. Harel *et al.* [2012] described that strawberry plants grown on biochar-amended substrate showed an upregulated salicylic acid-induced (SAR) and jasmonic acid/ethylene-induced systemic resistance (ISR) gene expression and were primed for gene expression upon infection by *Botrytis cinerea* and by *Podosphaera aphanis.* George *et al.* [2016] found that addition of spelt husk biochar and pine bark biochar to soil reduced infection rates of the root-lesion nematode *Practylenchus penetrans* and associated mass loss in carrot.

On the other hand, high sorption capacity of biochar may influence efficiency of plant protection with pesticides. Data from several studies have shown that BC reduces the efficacy of some pesticides due to bind-ing of their active substances and hence lowering their availability and activity towards the target organisms [Cox, 2012; Kookana, 2011]. In fact, only a small amount of sprayed or dusted pesticides fall directly on the target organisms. Most of them fall on the growing medium surface by direct precipitation or spray drift. It poses a severe threat to beneficial microbes, but also increases the risk of the pesticide residues being taken up by the plant roots and moved to other plant tissues, including the fruit. In such situations, particular green-waste biochars that have the ability of retaining many classes of plant protection products [Ćwieląg-Piasecka *et al.*, 2018], may be a great remedy to control the pesticide residue levels both in soils and in soilless beds.

According to the studies of [Ćwieląg-Piasecka *et al.*, 2018], con-ducted both on the pure wheat-straw BC and on the growing substrate where the BC was one of the components (data not published) — BC presence reduced the amount of pesticides availability to plants up to 99%. A very strong retention on BC-perlite substrate, where the cucumber and pepper were grown, was observed for agrochemicals of low water solubility (carbamates, chloroacetanilides). Another studied group of pes-ticides, namely phenoxyacetic acids (compounds of a great water solubil-ity; e.g. 2,4-D, MCPA), was less strongly retained by the pure BC (about 40% of the dose applied), but when it was mixed with the perlite, it increased to 70–95%. The addition of water to the growing medium with pesticides retained on their surface had no effect on their release, even after several cycles of rinsing. An important conclusion derived from the above experiments is that the utilization of a particular biochar in a greenhouse production can significantly reduce the overexposure of plants to pesticides and help to maintain the maximum residue levels (MRL) of plant protection products to a necessary minimum.

4.7 Biochar for Reduction of Environmental Impacts of Horticulture Production

Horticultural production is primarily involved in the intensive use of resources, such as land, water and inputs such as fertilizers and pesticides. Intensive horticultural practices over the twentieth century, coupled with growing greenhouse production, have dramatically impacted on the horticultural landscape. Land and soil degradation caused by erosion, organic matter decline, compaction, salinization, reduced fertility and pollution all have the potential for environmental mismanagement within horticulture. Maintaining and improving fertility is critical to increasing yield. Horticulture is water-consuming. As the majority of horticultural crops are irrigated, the industry contributes to water scarcity problems and carries a responsibility to reduce water use and causes of contamination. Intensive crop production requires higher levels of inorganic fertilizers. As horticultural crops are usually high value, the relative cost of these inputs to producers is important, and as a result they may be applied in excess, consequently increasing the risk of land and water pollution. Horticultural industry also releases high amounts of pesticides into the environment, however, nowadays there is an increasing awareness of the impact of these substances on human health.

Greenhouse crop production also has significant impact on CO_2 emissions, this is due to the combustion of coal and gas for heating. The contribution made by horticulture to global warming and climate change is through the processes of energy combustion, transportation, refrigerated storage and inorganic inputs. Finally, horticulture generates thousands of tonnes of green waste annually that need to be disposed.

Biochar seems to be a multitask tool in climate mitigation, bringing potential environmental benefits and reduction of negative footprints of crop production. Converting biomass into biochar locks carbon, which significantly increases the residence time of carbon in soil compared to the decomposition of raw biomass, where much of the carbon is converted to CO_2. Thus, sustainable biochar systems can be carbon negative.

The important result is a net reduction of CO_2 in the atmosphere. Methane and nitrous oxide are two potent greenhouse gases that are more effective than carbon at trapping heat inside earth's atmosphere. Methane is 25 times more powerful as a greenhouse gas, while nitrous oxide is 298 times more powerful [IPCC, 2007]. Biochar is a potential tool to mitigate the emissions of both of these greenhouse gases in the agricultural industry. Nitrous oxide emissions are controlled by a number of variables in agriculture. However, one major source of emissions are nitrogen fertilizers. Horticulture crops in soilless substrates require frequent additions of N, P and K fertilizers in high doses due to a little ability to supply or retain nutrients in amounts sufficient to sustain plants in small volume containers. Nitrates and phosphates from fertilizers are potential environmental hazards if they enter the groundwater or surface water by runoff or leaching. Most potting mixes are inert without application of fertilizers. Second, many irrigation practices, especially overhead watering with a hose, are very inefficient in terms of water and nutrient loss. Third, water-soluble fertilizers are often used at rates in excess of the plants' needs without regard for volume applied and frequency of application. Biochar can improve nutrient retention [Lehmann and Joseph, 2009], which decreases the need for fertilizers. The relationship between nitrogen fertilizer and nitrous oxide fluxes is linearly related, which means that nitrous oxide emissions could be reduced significantly from using biochar. A beneficial reduction in nutrient leaching has been observed in a multitude of biochar studies. The ability of biochar to reduce nutrient leaching is largely attributed to the charge and surface area properties of biochar. The increased nutrient retention and reduced leaching create obvious benefits to plant yield by increasing the bioavailability of essential nutrients, particularly in the presence of added components. Several previous investigations, focusing on soil and biochar mixtures, have shown that when biochar is added to soil, it can reduce the leaching of PO_4 −P, NO_3 −N and NH_4 −N, which therefore implies these nutrients are bound to biochar. Biochar amendments can alter the dynamics of N cycling, including improvements in N retention. Altland and Locke [2013] conducted a leaching experiment and they found that the addition of biochar to peat leads to a reduction of nitrate in comparison to a 100% peat bed. This might be effective in moderating the fluctuation of nitrate in a growing media when nitrate in liquid form is added. A number of studies have reported that adding biochar to soils may increase net nitrification rate and N mineralization [Zwieten *et al.*, 2010; Gul and Whalen, 2016; Kim *et al.*,

2017; Pereira *et al.*, 2017]. The enhancement of N bioavailability and N retention may result in potential agronomic and environmental benefits, such as reduction in fertilizer application and decline of the cost of food production.

Biochar as a component of growing media might be effective in reduced water consumption in greenhouse production. Horticulture needs good quality, toxic compounds and pathogen-free freshwaters facing scarcity problems and rising water costs. Good water management using well-designed systems is critical for sustaining production and quality of the product. Biochar seems to be a promising material in sustainable water management in horticulture production. In experiment with biochar–perlite soilless media in tomato and cucumber production, use of biochar addition leads to 20% and 25% reduction of irrigation only by increasing WHC of substrate. Average content of water in containers with biochar application during vegetation period was even 47% higher than in control on 100% perlite. High water retention and nearly 0% leaching from the container can benefit with reduced water use and protection of groundwater supplies, which are usually impacted by the leachates from horticultural irrigation.

Horticulture production brings risk of toxic substance transfer in the food chain due to the use of large amounts of fertilizers and pesticides. Water eutrophication and plant protection products' residues that can be found in waters and soils in the vicinity of greenhouses and horticultural landfills are an increasing problem worldwide. Fertilizers may contain different amounts of heavy metals, e.g. cadmium, lead, arsenic or zinc derived from mineral components, which can be easily dissolved during plant feed preparation and disposed in growing medium during fertigation procedures. These may lead to element accumulation in plants and reduce the quality of the product. High sorption capacity of biochar can also be important in plant protection and improvement of product quality, reducing the pollutant transfer to proximal waters or receptor organisms. Organic materials are a popular choice for this as they are derived from biological matter and often require little pre-treatment before they may be directly applied. For inorganic contaminants, which cannot be degraded by microbial action, such as heavy metals and some pesticides, reducing the bioavailability by sorption on biochar surface is one of the most effective strategies for minimizing risk of their transfer to plant tissues or preventing from leaching to ground waters.

Finally, as a stable and recalcitrant material, biochar can be reused in many vegetative cycles without loss of beneficial properties like water retention capacity. This may benefit with spare cost of substrate purchase. Sorption capacity for nutrient storage increases with time and biochar becomes more similar in properties as peat, serving as a "ready to use" source of available nutrients in plants. There is also an increasing interest in biochar reuse after horticulture production, e.g. in ornamental plant potting mixtures or in soil reclamation and fertility improvement.

All mentioned benefits of biochar use as a component or substitute of standard growing media materials make it "win–win" solution for horticulture in terms of adjustments for climate change and reduction of environmental impacts with the perspective of lowering the cost of production. Even though the cost of biochar is higher compared with standard growing media components, it can reimburse the expenses of water and fertilizer use, losses of crops caused by fungal diseases or pest affectations and, what is the most important, it has the potential to increase production yields.

References

Aller, D., Rathke, S., Laird, D., Cruse, R. and Hatfield, J. (2017). Impacts of fresh and aged biochars on plant available water and water use efficiency. *Geoderma*, 307, pp. 114–121.

Altland, J. E. and Locke, J. C. (2013). Effect of biochar type on macronutrient retention and release from soilless substrate. *HortScience*, 48, pp. 1397–1402.

Awad, Y. M., Lee, S. E., Ahmed, M. B. M., Vu, N. T., Farooq, M., Kim, I. S., Kim, H. S., Vithanage, M., Usman, A. R. A., Al-Wabel, M., Meers, E., Kwon, E. E., Ok, Y. S. (2017). Biochar, a potential hydroponic growth substrate, enhances the nutritional status and growth of leafy vegetables. *J. Clean. Prod.*, 156, pp. 581–588.

Barrett, G. E., Alexander, P. D., Robinson, J. S. and Bragg, N. C. (2016). Achieving environmentally sustainable growing media for soilless plant cultivation systems — A review. *Sci. Hortic. (Amsterdam)*, 212, pp. 220–234.

Blok, C., Regelink, I., Hofl, J. and Streminska, M. (2016). Perspectives for the use of biochar in horticulture. Report GTB-1388, pp. 1–43.

Buecker, J., Kloss, S., Wimmer, B., Rempt, F., Zehetner, F. and Soja, G. (2016). Leachate composition of temperate agricultural soils in response to biochar application. *Water Air Soil Pollut.*, 227, pp. 49–58.

Chan, K. Y., Van Zwieten, B. L., Meszaros, I., Downie, A. and Joseph, S. (2007). Agronomic values of greenwaste biochar as a soil amendment. *Aust. J. Soil Res.*, 45, pp. 629–634.

Conversa, G., Bonasia, A., Lazzizera, C. and Elia, A. (2015). Influence of biochar, mycorrhizal inoculation, and fertilizer rate on growth and flowering of Pelargonium (*Pelargonium zonale* L.) plants. *Front. Plant Sci.*, 6, pp. 429–438.

Cox, J. (2012). Biochar in horticulture: Prospects for the use of biochar in Australian horticulture. *NSW Dep. Prim. Ind.*, pp. 1–104.

Ćwieląg-Piasecka, I., Medyńska-Juraszek, A., Jerzykiewicz, M., Dębicka, M., Bekier, J., Jamroz, E. and Kawałko, D. (2018). Humic acid and biochar as specific sorbents of pesticides. *J. Soils Sediments*, pp. 1–11.

De Tender, C., Haegeman, A., Vandecasteele, B., Clement, L., Cremelie, P., Dawyndt, P., Maes, M. and Debode, J. (2016). Dynamics in the strawberry rhizosphere microbiome in response to biochar and Botrytis cinerea leaf infection. *Front. in Microbiol.*, 7, pp. 1–14.

Dumroese, R. K., Heiskanen, J., Englund, K. and Tervahauta, A. (2011). Pelleted biochar: Chemical and physical properties show potential use as a substrate in container nurseries. *Biomass Bioenergy*, 35, pp. 2018–2027.

Dunlop, S. J., Arbestain, M. C., Bishop, P. A. and Wargent, J. J. (2015). Closing the loop: Use of biochar produced from tomato crop green waste as a substrate for soilless, hydroponic tomato production. *HortScience*, 50, pp. 1572–1581.

Elad, Y., David, D. R., Harel, Y. M., Borenshtein, M., Kalifa, H. B., Silber, A. and Graber, E. R. (2010). Induction of systemic resistance in plants by biochar, a soil-applied carbon sequestering agent. *Phytopathology*, 100, pp. 913–921.

Elmer, W. H., Pathology, P., Pignatello, J. J. and Sciences, E. (2011). Effect of biochar amendments on mycorrhizal associations and fusarium crown and root rot of asparagus in replant soils. *Plant Disease*, 95, pp. 960–966.

Fidel, R. B., Laird, D. A., Thompson, M. L. and Lawrinenko, M. (2017). Characterization and quantification of biochar alkalinity. *Chemosphere*, 167, pp. 367–373.

Gascó, G., Cely, P., Paz-Ferreiro, J., Plaza, C. and Méndez, A. (2016). Relation between biochar properties and effects on seed germination and plant development. *Biol. Agric. Hortic.*, 32, pp. 237–247.

George, C., Kohler, J. and Rillig, M. C. (2016). Biochars reduce infection rates of the root-lesion nematode Pratylenchus penetrans and associated biomass loss in carrot. *Soil Biol. Biochem.*, 95, pp. 11–18.

Glaser, B., Balashov, E., Haumaier, L., Guggenberger, G. and Zech, W. (2000). Black carbon in density fractions of anthropogenic soils the Brazilian Amazon region. *Org. Geochem.*, 31, pp. 669–678.

Graber, E. R., Meller Harel, Y., Kolton, M., Cytryn, E., Silber, A., Rav David, D., Tsechansky, L., Borenshtein, M. and Elad, Y. (2010). Biochar impact on development and productivity of pepper and tomato grown in fertigated soilless media. *Plant Soil*, 337, pp. 481–496.

Graber, E. R., Tsechansky, L., Mayzlish-Gati, E., Shema, R. and Koltai, H. (2015). A humic substances product extracted from biochar reduces Arabidopsis root hair density and length under P-sufficient and P-starvation conditions. *Plant Soil*, 395, pp. 21–30.

Gray, M., Johnson, M. G., Dragila, M. I. and Kleber, M. (2014). Water uptake in biochars: The roles of porosity and hydrophobicity. *Biomass Bioenergy*, 61, pp. 196–205.

Güereña, D. T., Lehmann, J., Thies, J. E., Enders, A., Karanja, N. and Neufeldt, H. (2015). Partitioning the contributions of biochar properties to enhanced biological nitrogen fixation in common bean (*Phaseolus vulgaris*). *Biol. Fertil. Soils*, 51, pp. 479–491.

Harel, Y. M., Elad, Y., Rav-David, D., Borenstein, M., Shulchani, R., Lew, B. and Graber, E. R. (2012). Biochar mediates systemic response of strawberry to foliar fungal pathogens. *Plant Soil*, 357, pp. 245–257.

Hossain, M. K., Strezov Vladimir, V., Chan, K. Y., Ziolkowski, A. and Nelson, P. F. (2011). Influence of pyrolysis temperature on production and nutrient properties of wastewater sludge biochar. *J. Environ. Manage.*, 92, pp. 223–228.

IPCC, (2007). Climate Change 2007: Synthesis Report. Contribution of Working Groups I, II and III to the Fourth Assessment Report of the Intergovernmental Panel on Climate Change [Core Writing Team, Pachauri, R.K. and Reisinger, A. (eds.)]. IPCC, Geneva, Switzerland, p. 104.

Jiang, J., Yuan, M., Xu, R. and Bish, D. L. (2015). Mobilization of phosphate in variable-charge soils amended with biochars derived from crop straws. *Soil Tillage Res.*, 146, pp. 139–147.

Kadota, M. and Niimi, Y. (2004). Effects of charcoal with pyroligneous acid and barnyard manure on bedding plants. *Sci. Hortic. (Amsterdam)*, 101, pp. 327–332.

Kinney, T. J., Masiello, C. A., Dugan, B., Hockaday, W. C., Dean, M. R., Zygourakis, K. and Barnes, R. T. (2012). Hydrologic properties of biochars produced at different temperatures. *Biomass Bioenergy*, 41, pp. 34–43.

Kookana, R. S. (2011). Properties of biochar as affected by feedstock and production technology, In: Sparks, D.L. (ed.), *Advances in Agronomy*. Elsevier, pp. 108–109.

Laird, D. A., Fleming, P., Davis, D. D., Horton, R., Wang, B. and Karlen, D. L. (2010). Impact of biochar amendments on the quality of a typical Midwestern agricultural soil. *Geoderma*, 158, pp. 443–449.

Lehmann, J. and Joseph, S. (2009). *Biochar for Environmental Management: Science, Technology and Implementation*. London Sterling VA, pp. 1–405.

Lehmann, J., Rillig, M. C., Thies, J., Masiello, C. A., Hockaday, W. C. and Crowley, D. (2011). Biochar effects on soil biota — A review. *Soil Biol. Biochem.*, 43, pp. 1812–1836.

Locke, J., Altland, J. and Ford, C. (2013). Gasified rice hull biochar affects nutrition and growth of horticultural crops in container substrates. *J. Environ. Hortic.*, 31, pp. 195–202.

Margenot, A. J., Griffin, D. E., Alves, B. S. Q., Rippner, D. A., Li, C. and Parikh, S. J. (2018). Substitution of peat moss with softwood biochar for soil-free marigold growth. *Ind. Crops Prod.*, 112, pp. 160–169.

Méndez, A., Plaza, C. and Gascó, G. (2017). The effect of sewage sludge biochar on peat-based growing media. *Biol. Agric. Hortic.*, 33, pp. 40–51.

Mitra, J. and Raghu, K. (1998). Pesticides-non target plants interactions: An overview. *Arch. Agron. Soil Sci.*, 43, pp. 445–500.

Mollinedo, J., Schumacher, T. E. and Chintala, R. (2015). Influence of feedstocks and pyrolysis on biochar' s capacity to modify soil water retention characteristics. *J. Anal. Appl. Pyrolysis*, 114, pp. 100–108.

Novak, J. M., Busscher, W. J., Laird, D. L., Ahmedna, M., Watts, D. W. and Niandou, M. A. S. (2009). Impact of biochar amendment on fertility of a southeastern coastal plain soil. *Soil Sci.*, 174, pp. 105–112.

Park, J. H., Lamb, D., Paneerselvam, P., Choppala, G., Bolan, N. and Chung, J.-W. (2010). Role of organic amendments on enhanced bioremediation of heavy metal(loid) contaminated soils. *J. Hazard. Mater.*, 185, pp. 549–574.

Pereira, E. I. P., Conz, R. F. and Six, J. (2017). Nitrogen utilization and environmental losses in organic greenhouse lettuce amended with two distinct biochars. *Sci. Total Environ.*, 598, pp. 1169–1176.

Prasad, M., Tzortzakis, N. and McDaniel, N. (2017). Chemical characterization of biochar and assessment of the nutrient dynamics by means of preliminary plant growth tests. *J. Environ. Manage.*, 216, pp. 89–95.

Ruamrungsri, S., W. Bundithya, N. Potapohn, N. Ohtake, K. Sueyoshi and Ohyama, T. (2011). Effect of NPK levels on growth and bulb quality of some geophytes in substrate culture. *Acta Hort.*, 886, pp. 213–218.

Spokas, K. A., Novak, J. M., Stewart, C. E., Cantrell, K. B., Uchimiya, M., DuSaire, M. G. and Ro, K. S. (2011). Qualitative analysis of volatile organic compounds on biochar. *Chemosphere*, 85, pp. 869–882.

Spokas, K. A., Novak, J. M. and Venterea, R. T. (2012). Biochar's role as an alternative N-fertilizer: Ammonia capture. *Plant Soil*, 350, pp. 35–42.

Vaughn, S. F., Dan Dinelli, F., Jackson, M. A., Vaughan, M. M. and Peterson, S.C. (2018). Biochar-organic amendment mixtures added to simulated golf greens under reduced chemical fertilization increase creeping bentgrass growth. *Ind. Crops Prod.*, 111, pp. 667–672.

Vaughn, S. F., Eller, F. J., Evangelista, R. L., Moser, B. R., Lee, E., Wagner, R. E. and Peterson, S. C. (2015). Evaluation of biochar-anaerobic potato digestate mixtures as renewable components of horticultural potting media. *Ind. Crops Prod.*, 65, pp. 467–471.

Vaughn, S. F., Kenar, J. A., Thompson, A. R. and Peterson, S. C. (2013). Comparison of biochars derived from wood pellets and pelletized wheat straw as replacements for peat in potting substrates. *Ind. Crops Prod.*, 51, pp. 437–443.

Warnock, D. D., Mummey, D. L., McBride, B., Major, J., Lehmann, J. and Rillig, M. C. (2010). Influences of non-herbaceous biochar on arbuscular mycorrhizal fungal abundances in roots and soils: Results from growth-chamber and field experiments. *Appl. Soil Ecol.*, 46, pp. 450–456.

Zhang, J., Wang, Q., Feng, W., Shi, Z., Jiang, J. and Xia, J. (2016). Biochar addition reduced net N mineralization of a coastal wetland soil in the Yellow River Delta, China. *Soil Biol. Biochem.*, 288, pp. 329–336.

Zhang, L., Sun, X. Y., Tian, Y. and Gong, X. Q. (2014). Biochar and humic acid amendments improve the quality of composted green waste as a growth medium for the ornamental plant *Calathea insignis*. *Sci. Hortic. (Amsterdam)*, 176, pp. 70–78.

Zwart, D. C. and Kim, S. (2012). Biochar amendment increases resistance to stem lesions caused by phytophthora spp. in tree seedlings. *HortScience*, 47, pp. 1736–1740.

Zwieten, L. V., Kimber, S., Downie, A., Pyrolysis, P. and Morris, S. (2010). A glasshouse study on the interaction of low mineral ash biochar with nitrogen in a sandy soil. *Australian Journal of Soil Research*, 48, pp. 569–576.

Chapter 5

Biochar Application for Mine Land Reclamation: Metal Mining

Gabriel Gascó[*,§], Eliana Cárdenas-Aguiar[†], Jorge Paz-Ferreiro[‡] and Ana Méndez[¶]

[*]*School of Agricultural, Food and Biosystems Engineering, Technical University of Madrid, Madrid, Spain*

[†]*National Environmental Licensing Authority, Bogotá, Colombia*

[‡]*School of Engineering, RMIT University, Melbourne, Australia*

[¶]*Mines and Energy School, Technical University of Madrid, Madrid, Spain*

[§]*gabriel.gasco@upm.es*

Abstract

Phytoextraction is a promising technique that implies the use of hyper accumulator plants to reduce the heavy metal content in soils. Heavy metals can produce stress in plants, resulting in slow growth rate and loss in biomass production. Nevertheless, the biochar addition to heavy metal contaminated soil can produce the immobilization of heavy metals that together with the improvement of the physical and chemical properties of soil can produce a high plant growth and a better metal uptake. The aim of this chapter is to give a holistic approach about mine land reclamation with a specific review of the effects of biochar addition and the possibility to maximize metal phytoextraction. The influence of biochar in soil physicochemical properties, soil biota and soil heavy metal behavior are discussed in detail. Also, the chapter addresses the term phytomining as a concept to recover specific metals and provide them an economic value.

5.1 Introduction

Mining soils are included within anthropogenic soils, specifically Technosols, which comprise soils altered by industrial or artisanal human development [Blum *et al.*, 2018]. Coal and metal extraction produces changes in soil's physical and chemical properties [Shrestha and Lal, 2011; Liu *et al.*, 2018a]. In addition, these activities represent a threat to ecosystem due to the loss of vegetation, change in soil microbiota and, in general, the modification of natural balance [Zhang *et al.*, 2010]. The presence of acid soil pH values, low fertility, poor water-holding capacity, reduced soil structure, low soil quality, low organic matter content and soil erosion constitute negative impacts associated with mining activities [Puga *et al.*, 2016; Liu *et al.*, 2018a; Li *et al.*, 2019].

The high amount of mineral waste not only produces variations in soil properties but can also generate environmental problems on a larger scale, including disasters, diseases, loss of biodiversity and, under an economic approach, the loss of wealth [Xia and Cai, 2002; Wong, 2003]. The main problem of metal mining is related with the high concentration of metals that can be toxic for soil associated organisms (flora and fauna). Phytotoxicity can be mentioned as a negative impact and occurs when the heavy metal content exceeds the threshold values and affects plant growth and vegetable biomass production [Paz-Ferreiro *et al.*, 2014].

Some metals are essentials for biota, but their accumulation leads to several consequences: (i) altering crucial functional groups, (ii) displacing other metal ions and (iii) altering the composition of biological molecules [Li *et al.*, 2019]. Heavy metals are non-biodegradable elements and the speciation of these could change with time because they persist in the atomic form [Sitarz-Palczak and Kalembkiewicz, 2012].

The remediation of heavy metal contaminated soils is a complicated task because of the aforementioned characteristics. Chemical and physical remediation technologies are expensive and time-consuming [Zhang *et al.*, 2013]. Several methods have been used for the remediation of mining soils and could be divided into *ex situ* and *in situ* techniques. Surface capping, encapsulation, electrokinetics, vitrification, soil flushing, immobilization, phytoremediation and bioremediation are some *in situ* techniques. Whereas landfilling, soil washing, solidification and thermal vitrification are classified as *ex situ* techniques [Liu *et al.*, 2018b]. However, the cons in some of these technologies remain in the cost, the potential wastes generated after the remediation process and the impact to the environment. For this reason, technologies such as phytoextraction and the use of amendments appear as environmentally friendly methods to restore or decontaminate soils.

Organic amendments have been profusely used to reduce the transfer of contaminants and immobilize them within the soil matrix. Biochar by itself has a number of properties that improve soil quality. In addition, the production of biochar is a useful tool in the management of wastes (i.e. manure waste, crop residues, industrial waste) as the pyrolysis process reduces the waste volume and facilitates the transportation of the materials.

Phytoextraction is a promising technique that implies the use of hyper accumulator plants to reduce the heavy metal content in soils [Li *et al.*, 2018; Dhiman *et al.*, 2016]. This approach requires a high growth rate and the ability to adapt to high heavy metal concentrations without reducing the plant biomass. However, heavy metal contaminated soils generate plant stress and as a result hyperaccumulators show slow growth rate and low biomass production [Paz-Ferreiro *et al.*, 2014]. In this case, the abovementioned addition of biochar to mining soils can be used as a strategy to overcome the limitation to plant growth, due to several changes in soil properties: (i) pH variations (liming effect) improve phytoremediation via reduction in Al toxicity and increasing available nutrients, (ii) increase of organic matter content can have a positive effect in biomass production, (iii) immobilization of heavy metals by biochar improves plant growth [Lu *et al.*, 2014]. Therefore, combining phytoextraction and biochar allows for high plant growth and a better metal uptake [Lu *et al.*, 2014, 2015; Rodríguez-Vila *et al.*, 2014].

5.2 Biochar for Mine Land Reclamation

Biochar properties are dependent on the biomass used in the pyrolytic process and the production conditions (Fig. 5.1). These variables affect the physicochemical properties of biochar and, consequently, produce changes in soil properties modifying the heavy metal behavior in soils.

5.2.1 *Influence of biochar on soil physicochemical properties*

Biochar addition produces changes in soil physicochemical properties involved in mine restoration: pH [Houben *et al.*, 2013a; Beesly *et al.*,

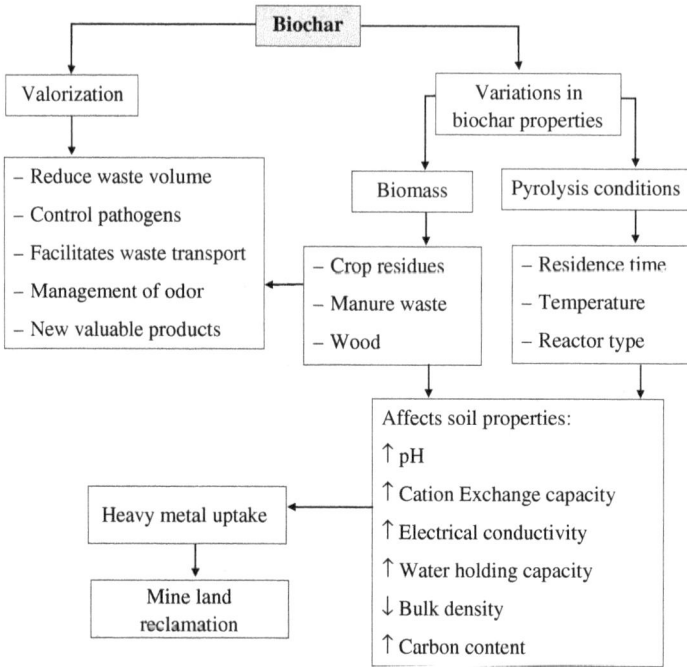

Figure 5.1:   Biochar implications for mine land reclamation.

2010], electrical conductivity (EC) [Chan *et al.*, 2008; Ahmad *et al.*, 2017; Burrell *et al.*, 2016], cation exchange capacity (CEC) [Jien and Wang, 2013], water holding capacity (WHC) [Herath *et al.*, 2013; Novak *et al.*, 2012], bulk density [Alburquerque, *et al.*, 2014; Burrell *et al.*, 2016; Laird *et al.*, 2010], porosity [Herath *et al.*, 2013; Karhu *et al.*, 2011], surface area (SA) [Tang *et al.*, 2013; Beesley *et al.*, 2011] and nutrient content [Prendergast-Miller *et al.*, 2014].

Regarding soil pH, biochar addition increases pH value, but the effect depends on the type of amended soil. This generalized trend could be applied for acidic soils. Houben *et al.* [2013a] found that the addition of Miscanthus straw derived biochar increases soil pH values even at the lowest proportion of this material due to the alkali nature of this material and the transformation of Ca, Mg, K and Na into oxides, hydroxides and carbonates, respectively. Beesly *et al.* [2010] also found changes in pH values in acidic soil (5.45) after the addition of hardwood-derived biochar (pH = 7.56). However, in the cases of soils with basic pH, the liming effect of biochar can lead to extremely high pH values [Alburquerque, *et al.*, 2014].

Additionally, biochar can modify soil EC. According to Chan *et al.* [2008], the use of two biochars from poultry litter pyrolyzed at 450°C and 550°C in an Alfisol changed the EC from 0.11–0.13 dS m^{-1} to 0.28–0.29 dS m^{-1} after the addition of 50 t ha^{-1} of biochar. An investigation made with biochars from different feedstock (soybean stove, peanut shells and pine needles) showed that in all cases the use of these amendments increases soil EC. The highest EC (0.40 dS m^{-1}) was obtained with soybean stove biochar (pyrolysis temperature: 700°C) added to an agricultural soil. Besides, they found a positive correlation between pH and EC [Ahmad *et al.*, 2017].

Burrell *et al.* [2016] explained, after a 3-year study, that initial EC increased with the addition of all types of biochars in a Planosol, a Chernozem and a Cambisol. Nevertheless, the EC did not show consistent behavior at the end of the experiment; only for the Planosol and the Cambisol an increase in EC values was noticed with the addition of woodchip and vineyard biochars prepared at 400°C.

The ability of biochar to improve CEC is a result of the amount of nutrients that are a direct supplier of exchangeable cations in soils [Lehmann *et al.*, 2003]. The increase of CEC in amended soils was established by Jien and Wang [2013]. These authors reported a CEC of 7.41 cmol$_{(+)}$ kg^{-1} for control soil with a variation up to 9.26 cmol$_{(+)}$ kg^{-1} with 2.5% of wood biochar addition and up to 10.8 cmol$_{(+)}$ kg^{-1} with a biochar rate of 5%. Slow oxidation process was the main mechanism involved with increase of carboxylic groups. The abovementioned results suggest better soil fertility and nutrient retention.

The determination of water holding capacity (WHC) in a 295 day experiment proved the efficiency of biochar addition even over a long-term period [Herath *et al.*, 2013]. The increase in hydrophilicity caused by oxidation of biochar surfaces leads to improved WHC. The same results were found with the addition of peanut hull, pecan shell, poultry litter and switch grass biochars in silt loam and loamy sand soils where the addition of 2% biochar improved WHC values. For example, after the application of a peanut hull biochar, produced at 400°C, to a loamy sand soil, the water holding capacity changed from 2.87 g cm^{-3} up to 3.97 g cm^{-3} in 28 days [Novak *et al.*, 2012].

There is a strong relationship between water retention and bulk density. Alburquerque *et al.* [2014] suggested the presence of low bulk density levels in a Haplic Luvisol amended with biochar made from biomass (olive stone, almond shell, pine woodchips, etc.). The data found in this study implied the presence of more pore space for water retention due to a decrease in bulk density. Similarly, Laird *et al.* [2010] explained the benefits of biochar addition in soil bulk density, indicating the function of this material as an effective soil conditioner. A large body of literature has been published on the effects of biochar in the bulk density of amended soils (Table 5.1).

Regarding soil porosity, there is an overall increase with biochar addition related to low bulk density values [Sun *et al.*, 2013]. The soil physical quality usually is related with higher porosity due to the involved functions of micropores and macropores. Micropores play an important role in

Table 5.1:   Bulk density for different soils amended with biochar.

Treatment	Bulk density (g cm^{-3})	References
Soil (Typic Fragiaqualf)	1.01	Herath *et al.* [2013]
Soil/corn stover (CS)	0.93	
Soil/CS biochar 350°C	0.94	
Soil/CS biochar 550°C	0.91	
Soil (Typic Paleudults)	1.42	Jien and Wang [2013][a]
Soil/wood biochar 700°C, 2.5%	1.15	
Soil/wood biochar 700°C, 5%	1.08	
Soil (Planosol)	1.35	Burrell *et al.* [2016][b]
Soil/wheat straw biochar 525°C	1.28	
Soil/woodchip biochar 525°C	1.32	
Soil/vineyard biochar 525°C	1.31	
Soil/vineyard biochar 400°C	1.25	
Soil (chernozem)	1.28	
Soil/woodchip biochar 525°C	1.23	
Soil (cambisol)	1.17	
Soil/woodchip biochar 525°C	1.11	
Soil	1.30	Karhu *et al.* [2011]
Soil/biochar of birch charcoal	1.25	

Notes: [a]Biochar of white lead trees wood.
[b]This is a 3-year experiment with data of bulk density after this time.

molecular adsorption and transport, while the main functions of macro-pores are related with aeration and water movement [Mukherjee and Lal, 2013].

Hseu *et al.* [2014] explained that the changes in soil porosity after biochar addition are related to soil structure. They established that the reorganization of soil particles and the formation of macropores lead to an increase in soil porosity. Jien and Wang [2013] reported the usefulness of biochar as a binding agent to improve the soil microaggregates connection, ending with the transformation to stable macroaggregates, crucial factors in order to preserve soil porosity. This parameter could be crucial considering the decline in soil erosion associated with biochar addition.

The high biochar total porosity influences soil WHC, with a rise of this soil property because of the retention of water molecules in small pores and the water flow through the larger pores [Karhu *et al.*, 2011]. In the same way, Herath *et al.* [2013] enlightened the influence of porosity in available water content, readily available water content, hydraulic conductivity and soil aeration.

Figure 5.2 represents the variation of soil porosity with (black bar) and without (gray bar) biochar addition for some investigations with two

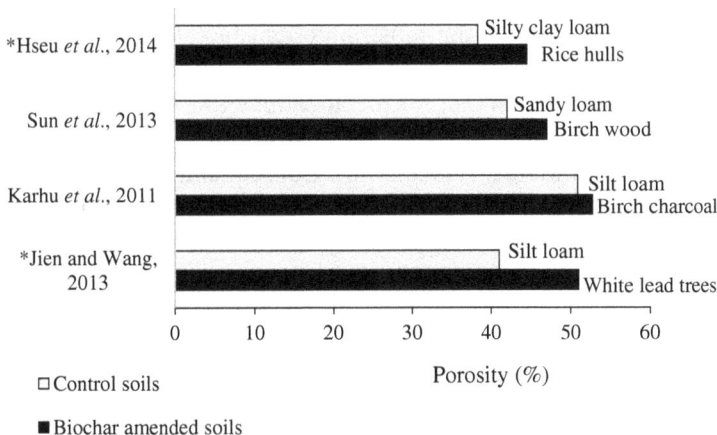

Figure 5.2:   Porosity (%) of biochar amended soils.

Note: * The values correspond to the minimum rate of biochar applied.

important specifications: type of soil and biochar feedstock. In general, the results show a positive effect of biochar addition in soil porosity, leading to a significant increase in the percentage of this parameter, in some cases even at the lowest rate of biochar addition (2.5%).

Also, the addition of biochar can modify the soil SA. For example, the addition of mixed hardwood derived biochar increased the soil SA of a fine loamy soil ($130 \, m^2 \, g^{-1}$) with a rise up to 133, 138 and $153 \, m^2 \, g^{-1}$ after the addition of different biochar rates, respectively (0.5%, 1% and 2%) [Laird *et al.*, 2010]. Tang *et al.* [2013] illustrated the changes of this property for different feedstocks and temperatures of biochar production. At higher temperatures, the biochar SA increases, generally leading to a higher soil SA after the use of this amendment.

Another implication of biochar addition is the shift of the biochar surface from a hydrophobic form to a hydrophilic one when mixed with the soil. The obtained result implies an increase in CEC, nutrient content and WHC, and in an extended way, improved soil health [Kambo and Dutta, 2015].

5.2.2 *Influence of biochar in soil heavy metal content*

The variation in the soil properties due to biochar addition affects in a direct way metal mobility, its bioavailability and its total content. Consequently, biochar has an important role in the treatment of mining soils, which are characterized by high heavy metal concentrations.

The soil pH is one of the most important factors in available metal content in soil, and the influence of biochar in soil pH warrants an effect in this property. For example, for metal cations like Cd, Zn and Pb, acid pH values increase the mobility, meanwhile, metal anions produce a decrease in metal mobility (As and Sb). Consequently, the rise of soil pH after biochar addition to alkaline or neutral values can produce the increase of available As and Sb mobility and the decrease of the available Cd, Zn and Pb [Bolan *et al.*, 2014; Lomaglio *et al.*, 2016].

Some investigations report that for As, the release and retention of oxide and hydroxide forms of Fe, Mn and Al in the biochar surface control the As amount. But more specifically, the retention of this metal is a result of Ca^{2+}, Na^+ and Mg^{2+} presence. All these changes are related to pH [de Brouwere *et al.*, 2004; Moreno-Jiménez *et al.*, 2012; Smith *et al.*, 2002; Stachowicz *et al.*, 2008]. On the contrary, the liming effect of biochar alters the mobility of Pb with the precipitation of metals' phosphates/carbonates that immobilize the metal in soil matrix. The same behavior as Pb has been reported for Cd and Zn [Lomaglio *et al.*, 2016].

For Zn, Zhang *et al.* [2017] reported a higher amount of the metal in the exchangeable fraction in an acid soil, but the soil pH variation after biochar addition, into a more alkaline pH, allowed the adsorption of Zn on biochar particles and the consequent equilibrium in dissolution and precipitation reactions. Puga *et al.* [2016] found the same trend of metals' reduction for Cd, Pb and Zn after the biochar liming effect. In the case of Cu, Beesley *et al.* [2010] reported similar results compared with as content, where Cu content increased with basic pH.

Biochar's role as an immobilizing amendment is due to different changes in soil properties that can affect metal speciation. Not only does pH affect the metal behavior but also the changes in CEC, WHC and SA [Kambo and Dutta, 2015]. For example, high SA and CEC facilitate metal retention, precipitation and coprecipitation. Therefore, all the approaches are connected because biochar addition produces changes in soil properties and these variations affect the metals' behavior.

A vast amount of information has been published about the decrease of available metal content after biochar addition. Park *et al.* [2013] observed a reduction of bioavailable Pb and Cd by sorption in two soils (spiked soil and natural contaminated soil) after the application of chicken manure and green waste biochars. The percentages of immobilization were 93.6% and 100% for Pb in the case of chicken manure biochar for spiked and natural soil, respectively. For Cd, with the same biochar the values were of 88.9% for the spiked soil and 94.6% for the natural one. In the case of green waste biochar, they found lower values of 36.8% and 30.3% for Cd and Pb for spiked soil and, in the natural contaminated soil, for Pb it was 72.9% and 30.3% for Cd. Beesley and Marmiroli [2011] concluded, after the application of a hardwood biochar in a column leaching experiment, that the main process involved in the reduction of soluble Cd and Zn content was sorption. Table 5.2 contains some investigations about heavy metal mobility, bioavailability and total content in soils amended with biochar.

The interaction between metals and biochar can vary for each metal [Kuppusamy *et al.*, 2016]. In conclusion, the remediation of heavy metal contaminated soils by biochar depends on several factors: (i) type of amended soil, (ii) biochar feedstock, (iii) pyrolysis temperature used in biochar production, (iv) target metal and (v) biochar rate.

Table 5.2:   Effect of biochar amended soils in the decrease of heavy metal concentration.

Effect	Feedstock	Metal	References
Mobility	Urban organic waste	Cu	Cardenas-Aguiar *et al.* [2017]
	Wood	Cd, Zn	Debela *et al.* [2012]
	Oak wood	Cu, Pb	Karami *et al.* [2011]
	Dairy manure	Pb	Cao *et al.* [2011]
	Hardwood	As, Cd, Cu, Pb, Zn	Beesley and Dickinson [2011]
	Hardwood	Cd, Zn	Beesley and Marmiroli [2011]
	Hardwood	Cd, Zn	Beesley *et al.* [2010]
	Broiler litter and pecan shells	Cu, Cd, Ni	Uchimiya *et al.* [2010]
	Bamboo	Cd	Ma *et al.* [2007]
Bioavailability	Willow	Cd, Pb, Cu, Zn	Gondek *et al.* [2018]
	Bamboo	Cd, Pb	Xu *et al.* [2016]
	Sugar cane straw	Zn, Pb, Cd	Puga *et al.* [2015]
	Wheat straw	Cd, Pb	Bian *et al.* [2014]
	Sewage sludge	As, Cr, Co, Cu, Ni, Pb	Khan *et al.* [2013]
	Miscanthus	Cd, Pb, Zn	Houben *et al.* [2013a]
	Chicken manure and green waste	Pb, Zn	Park *et al.* [2013]
	Chicken manure	Cr	Choppala *et al.* [2012]
	Sewage sludge	Cu, Ni, Zn, Cd, Pb	Méndez *et al.* [2012]
	Rice straw	Cu, Pb, Cd	Jiang *et al.* [2012]
	Quail litter	Cd	Suppadit *et al.* [2012]
	Oak wood	Pb	Ahmad *et al.* [2012]
	Chicken manure and green waste	Cd, Cu, Pb	Park *et al.* [2011]
	Orchard prune residue	Cd, Cr, Cu, Ni, Pb, Zn	Fellet *et al.* [2011]
	Eucalyptus	As, Cd, Cu, Pb, Zn	Namgay *et al.* [2010]
	Hardwood	As	Hartley *et al.* [2009]
	Cotton stalks	Cd	Zhou *et al.* [2008]

Table 5.2: (*Continued*)

Effect	Feedstock	Metal	References
Total content (mobility and bioavailability)	Not specified	Cu, Zn, Pb, Cd, Mn	Li *et al.* [2019]
	Rice straw	As	Chen *et al.* [2018]
	Magnetic biochars	Cd, Pb, Zn	Lu *et al.* [2018]
	Rice straw	Zn	Zhang *et al.* [2017]
	Wood-commercial charcoal	Cd, Zn, Pb	Lomaglio *et al.* [2016]
	Deinking paper sludge	Ni	Méndez *et al.* [2014]

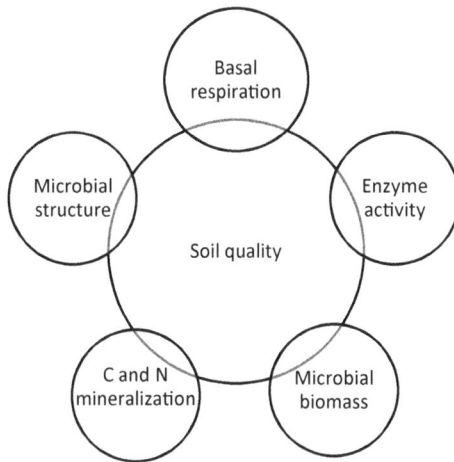

Figure 5.3:   Soil microbiological parameters used as soil quality indicators.

5.2.3 *Influence of heavy metals on soil biota*

Soil biota is a key factor to denote the impact of contaminants in this eco-system. Due to the presence of animals, plants and microorganisms the biological activities (i.e. nutrient cycling, organic matter decomposition, ecosystem sustainability, bioturbation, etc.) can maintain a state of equilibrium. Soil microorganisms are more sensitive to contamination and for that reason are essential for environmental management as indicators of soil quality (Fig. 5.3) [Zhang *et al.*, 2010].

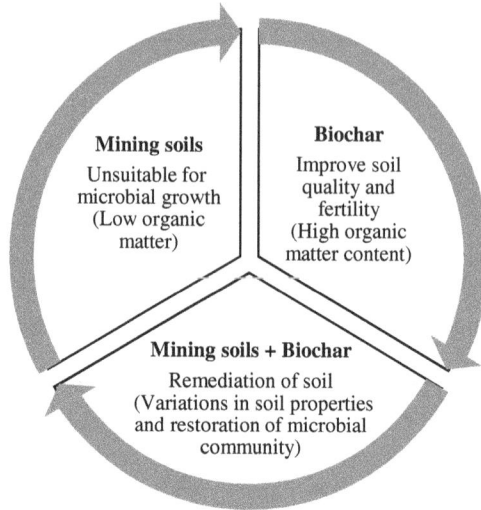

Figure 5.4:   Conceptual outline: Effects of mining soils and biochar addition on soils microorganisms.

Mining soils can have many problems due to the negative impacts on soil chemical, physical and biological properties as aforementioned. These variations produce an unsuitable environment for microbial growth. Low organic matter content and poor nutrient turnover in heavy metal contaminated soils are some of the limitations for soil biota [Novak *et al.*, 2018]. Also, pH changes can affect the microbial community, as a consequence of microorganism's preferences. Organic amendments to improve the soil quality after the exposure to high amount of metals are widely used. These materials contribute to an increase in soil organic matter and, subsequently, recover the soil properties, including: nutrient content and structure, and reduce the availability of toxic metals, stimulating microbial populations [Zornoza *et al.*, 2016]. In this case, the addition of biochar in mining soils has a synergic effect (Fig. 5.4) where the variations in soil properties and the alterations of metals' speciation cause modifications in microbial community. The different aspects of soils amended with biochar are explained in the following section.

5.2.4 *Influence of biochar on soil biota*

Biochar addition can affect soil microbiota in several ways (Table 5.3). Simultaneously, the responses of soil microorganisms can differ widely.

Table 5.3:   Changes in soil microorganisms after biochar addition.

Biochar addition	Consequences in soil microorganisms
↑ Nutrient and carbon availability	↑ Abundance of non-symbiotic microorganisms. No effects in symbiont organisms
pH up to 7 pH > 7	↑ Bacteria biomass. No changes in fungi biomass. ↓ Fungi biomass
↑ Bacterial adhesion to biochar	↑ Bacteria abundance. No effect of fungal abundance
↑ Sorption of toxic compounds	↑ Microbial abundance
↑ C/N ratio	↑ Fungi biomass
↑ Water held in biochar pores	↑ Microbial biomass
↑ SA ↑ WHC ↑ hydration	↑ Bacteria growth
Protection from grazers	↑ Fungi and bacteria biomass
Biofilm formation	↑ Bacteria growth

Sources: Lehmann *et al.* [2011], Chan *et al.* [2008], Gregory *et al.* [2014] and Paz-Ferreiro *et al.* [2012].

The effect depends on the existing nutrient and C availability in soil, the magnitude and direction of change, the microorganism group and pre-existing soil pH. In the case of bacterial adhesion to biochar, the biochar properties have an important influence (i.e. ash content, pore size, volatile matter content). Biochar addition also increases the diversity of microbial community and results in richness from a taxonomic point of view (presence of new clades) [Lehmann *et al.*, 2011].

Microorganisms can be disturbed by soil heavy metal contamination, as part of the consequences of metal mining activities, but the addition of biochar can improve this negative impact and recover the soil microbiota structure. These responses have been widely investigated by several authors [Chan *et al.*, 2008; Gregory *et al.*, 2014; Paz-Ferreiro *et al.*, 2012].

5.2.5 *Biochar combining with phytoextraction*

The techniques of phytoremediation include phytofiltration, phytostabilization, phytovolatilization, phytodegradation, rhizodegradation, phytodesalination and phytoextraction [Paz-Ferreiro *et al.*, 2014]. Phytoextraction of heavy metals occurs in the roots and then the plants translocate the metal to the shoots (stems and leaves). Plant potential to develop this kind of process depends on shoot metal concentration and shoot biomass.

Metal uptake by plants has been verified in two ways that involve the use of hyperaccumulator plants with high and low aboveground biomass production (Fig. 5.5) [Ali *et al.*, 2013].

According to Bhargava *et al.* [2012] a vast number of plant species have the potential to accumulate heavy metals (450 species). Figure 5.6

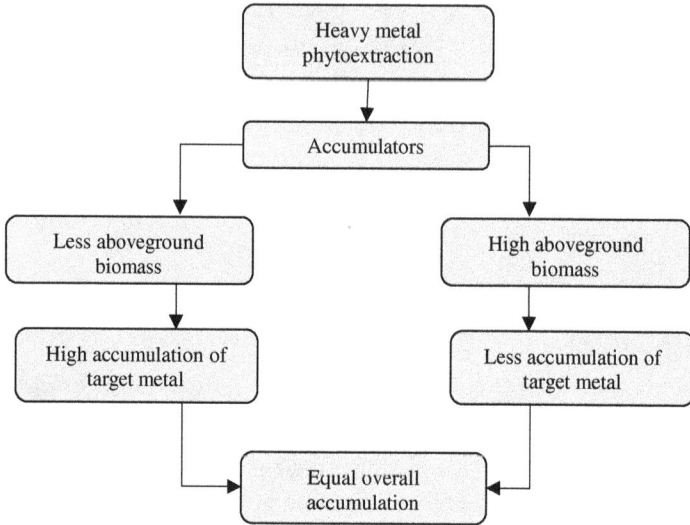

Figure 5.5:   Approaches of heavy metal phytoextraction.

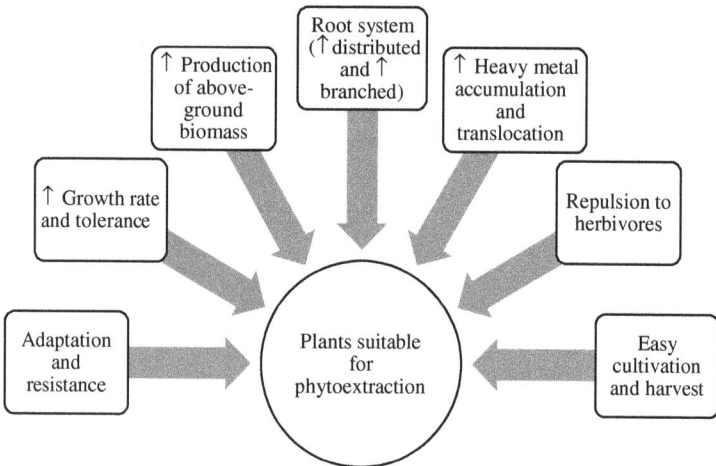

Figure 5.6:   Main characteristics of plants suitable for phytoextraction.

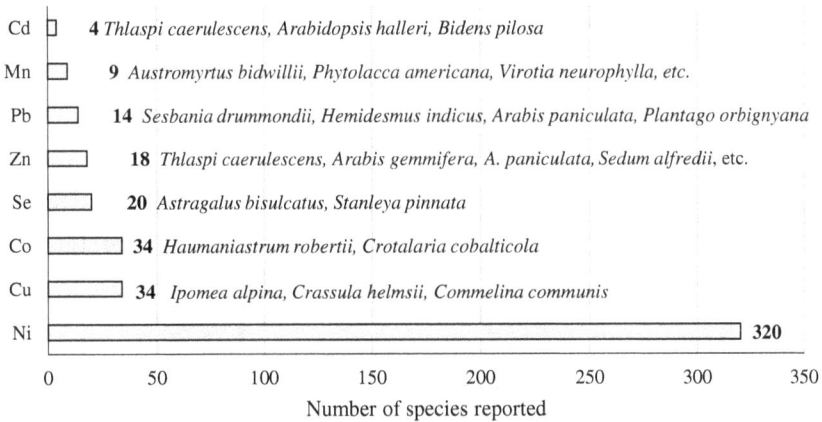

Figure 5.7:   Plant species that are heavy metal hyperaccumulators.

Source: Adapted from Bhargava *et al.* [2012].

shows the main characteristics of plants suitable for phytoextraction [Paz-Ferreiro *et al.*, 2014]. The use of one specific plant relies on the target metal. The Ni extraction is the most investigated one, with approximately 320 plant species, some of them being: *Berkheya coddii, Alyssum serpyl-lifolium, Alyssum bertolonii, Sebertia acuminata, Phidiasialindavii, Bornmuellera kiyakii.* Others metals have less hyperaccumulators reported. Figure 5.7 demonstrates the number of species identified for phytoextraction of each metal and some examples of these, except for Ni, as they have been mentioned above [Bhargava *et al.*, 2012]. Recently, investigations reported the important extraction role of *Brassica napus* and *Sinapis alba* for metals, including: Cd, Zn, Pb and Cu [Dhiman *et al.*, 2016; Houben *et al.*, 2013b; Kloss *et al.*, 2014].

Heavy metal contaminated soils can produce toxicity in plants, followed by signs of stress. The symptoms could be the presence of leaf chlorosis, dehydration and growth delay [Houben *et al.*, 2013b]. Also, factors like soil cation exchange capacity, pH, organic matter content and oxidation metal state can influence the plant growth and extraction process. Given this information, biochar addition also affects the phytoextraction of heavy metals in mining soils due to the influence in soils' properties. Additionally, it is important to highlight the disadvantages of phytoextraction: (i) long period is required for remediation, (ii) few target metals,

Table 5.4:   Combining phytoextraction and the use of biochar.

Disadvantages of phytoextraction	Improvement with biochar addition
Unsuitable in areas where the metal concentration is too elevated: Phytotoxicity	Biochar can reduce metal mobility and bioavailability leading to a better plant growth: ↓ Phytotoxicity.
Plant stress due to low available nutrient content	Soil nutrient content and organic matter increase with biochar addition, so biochar can reduce plant stress, enhancing soil fertility.
Presence of acid soils pH disturb plant growth	Biochar liming effect improves pH and allows efficient plant growth.
Sensibility to pathogens	Biochar changes microbial community leading to the presence of organisms that generate antibiotics and protect plants from pathogens.
Slow growth rate and low biomass production	High plant yields after biochar are accredited to augmented water and nutrient retention, improved biological properties and CEC and effects on nutrient cycling. Also biochar acts as a slow liberation fertilizer that increases biomass production.

(iii) roots are not able to reach deep soil horizons, and (iv) absence of knowledge of the agronomic properties and management, among others [Sarwar *et al.*, 2017]. However, some of the weaknesses of this technique can be overcome by biochar addition maximizing metal uptake. Table 5.4 shows the disadvantages of phytoremediation and the role of biochar as a preliminary step to mine land remediation to enhance phytoextraction.

The aim of this chapter was to give a holistic approach about mine land reclamation, with a specific review of biochar addition and the changes in soil properties and the possibility to maximize metal phytoextraction (Fig. 5.8). Table 5.5 contains some investigations of biochar addition and the combination with phytoextraction to remediate heavy metal contaminated soils.

New perspectives have emerged and been related to the phytoextraction technique. Currently the term "phytomining" is widely used as a modern concept to recover specific metals and provide them an economic value. This implies the use of plants to accumulate metals in the

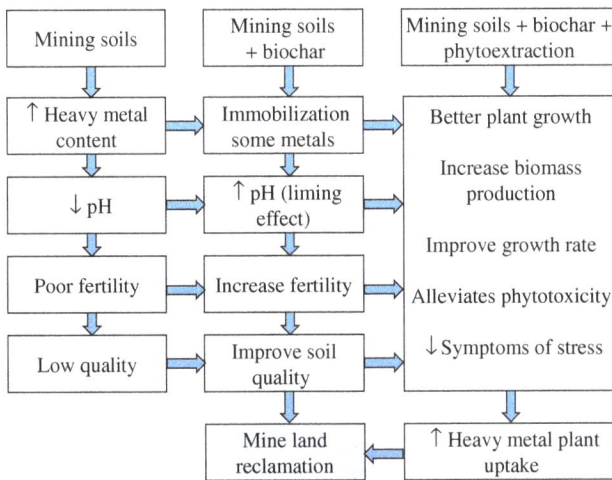

Figure 5.8: Holistic approach about mine land reclamation with biochar and phyto-extraction.

harvestable biomass and then extraction of the metal specie or compounds. A suitable variant of phytomining could be "agromining", which involves an alternative type of agriculture on contaminated lands [Van der Ent *et al.*, 2015; Rue *et al.*, 2019].

5.2.6 *Agromining*

Agromining involves several steps to obtain the target metal. The first step would be the growth of the plant, then the harvesting biomass, followed by thermal process (drying and ashing) and finally the recovery of the compound. This production sequence generates oxides, salts and complexes, which are metallic compounds (Fig. 5.9). Possible scenarios for the use of this technique could be degraded mined land or low-productivity agricultural soils. With the approach aforementioned, a new income could be generated for local communities where the impact of degraded lands affect the economic wealth [Van der Ent *et al.*, 2015; Rue *et al.*, 2019]. Also in this case, biochar could improve the obtained results due to effects like: higher biomass production, better plant growth, decrease of plant stress, among other beneficial impacts (Table 5.4) [Rue *et al.*, 2019].

Table 5.5:   Combining phytoextraction and biochar to remediate metal contaminated soils.

Biochar	Soil type	Metal	Plant species	References
Urban organic waste	Typic xerofluvent	Cu	*Lepidium sativum Sinapis alba Lolium* sp.	Cardenas–Aguiar et al. [2017]
Eucalyptus and poultry litter	Fimic anthrosol	Cd	*Amaranthus tricolor*	Lu et al. [2015]
Willow *Salix* sp	Pipe clay	As	*Lolium perenne*	Gregory et al. [2014]
Miscanthus straw	Sandy loam	Cd, Zn, Pb	*Brassica napus* L.	Houben et al. [2013b]
Corn straw	Typic agri-udic ferrosol	Cd, Zn	*Sedum plumbizincicola*	Li et al. [2018]
Holm oak wood	Mining soil	Co, Cu, Ni. Pb, Zn	*Brassica juncea* L.	Rodriguez-Vila et al. [2015]
Woodchip	Planosol, cambisol, chernozem	Al, Cd, Cu. Pb, Mn, As, B, Mo, Se	*Sinapis alba* L.	Kloss et al. [2014]
Mix of hardwood and softwood	Redoxic cambisols	Cd, Zn, Pb	*Noccaea caerulescens*	Rees et al. [2016]
Mix of hardwoods	3 soils	As	*Miscanthus x giganteus*	Hartley et al. [2009]
Pruning residues from orchards, fir tree pellets, manure pellets mixed with fir tree pellets	Technosol	Cd, Cr, Cu, Fe, Ni, Pb, Tl and Zn	*Anthyllis vulneraria subsp. polyphylla* (Dc.) Nyman, *Noccaea rotundifolium* (L.) Moench subsp. cepaeifolium and *Poa alpina* L. subsp. alpina	Fellet et al. [2014]

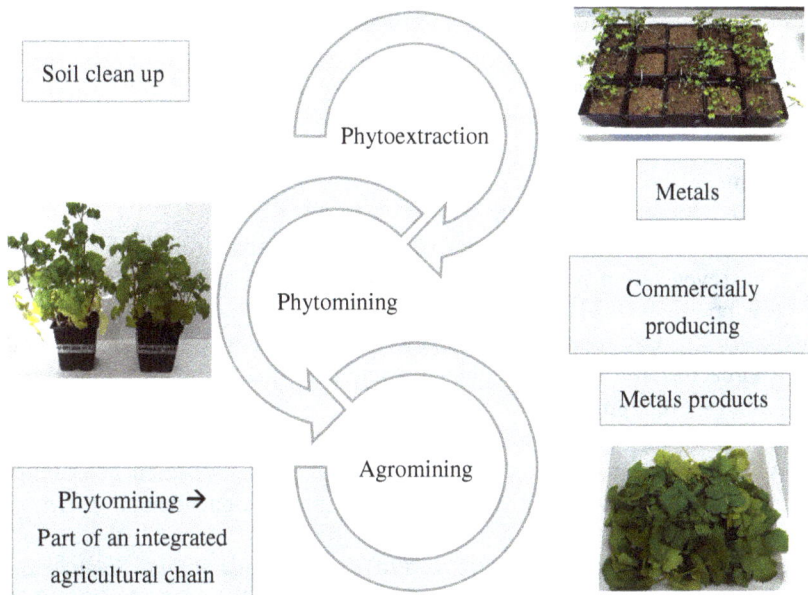

Figure 5.9:   Agromining: A new perspective with economic value.

References

Ahmad, M., Lee, S. S., Yang, J. E., Ro, H. M., Lee, Y. H. and Ok, Y. S. (2012). Effects of soil dilution and amendments (mussel shell, cow bone, and biochar) on Pb availability and phytotoxicity in military shooting range soil. *Ecotoxicol. Environ. Saf.*, 79, pp. 225–231.

Ahmad, M., Lee, S. S., Lee, S. E., Al-Wabel, M. I., Tsang, D. C. and Ok, Y. S. (2017). Biochar-induced changes in soil properties affected immobilization/mobilization of metals/metalloids in contaminated soils. *J. Soils Sediments*, 17(3), pp. 717–730.

Alburquerque, J. A., Calero, J. M., Barrón, V., Torrent, J., del Campillo, M. C., Gallardo, A. and Villar, R. (2014). Effects of biochars produced from different feedstocks on soil properties and sunflower growth. *J. Plant Nutr. Soil Sci.*, 177(1), pp. 16–25.

Ali, H., Khan, E. and Sajad, M. A. (2013). Phytoremediation of heavy metals — concepts and applications. *Chemosphere*, 91(7), pp. 869–881.

Beesley, L., Moreno-Jiménez, E. and Gomez-Eyles, J. L. (2010). Effects of biochar and greenwaste compost amendments on mobility, bioavailability and toxicity of inorganic and organic contaminants in a multi-element polluted soil. *Environ. Pollut.*, 158(6), pp. 2282–2287.

Beesley, L. and Dickinson, N. (2011). Carbon and trace element fluxes in the pore water of an urban soil following greenwaste compost, woody and biochar amendments, inoculated with the earthworm. *Lumbricus terrestris*. *Soil Biol. Biochem.*, 43(1), pp. 188–196.

Beesley, L. and Marmiroli, M. (2011). The immobilisation and retention of soluble arsenic, cadmium and zinc by biochar. *Environ. Pollut.*, 159, pp. 474–480.

Beesley, L., Moreno-Jiménez, E., Gomez-Eyles, J. L., Harris, E., Robinson, B. and Sizmur, T. (2011). A review of biochars' potential role in the remediation, revegetation and restoration of contaminated soils. *Environ. Pollut.*, 159, pp. 3269–3282.

Bhargava, A., Carmona, F. F., Bhargava, M. and Srivastava, S. (2012). Approaches for enhanced phytoextraction of heavy metals. *J. Environ. Manage.*, 105, pp. 103–120.

Bian, R., Joseph, S., Cui, L., Pan, G., Li, L., Liu, X. and Marjo, C. (2014). A three-year experiment confirms continuous immobilization of cadmium and lead in contaminated paddy field with biochar amendment. *J. Hazard. Mater.*, 272, pp. 121–128.

Blum, E. H., Schad, P. and Nortcliff, S. (2018). *Essentials of Soils Science: Soil Formation, Functions, Use and Classification* (World Reference Base, WRB) (Borntraeger Science Publishers, Stuttgart).

Bolan, N., Kunhikrishnan, A., Thangarajan, R., Kumpiene, J., Park, J., Makino, T., Kirkham, M. B. and Scheckel, K. (2014). Remediation of heavy metal (loid)s contaminated soils — To mobilize or to immobilize? *J. Hazard. Mater.*, 266, pp. 141–166.

Burrell, L. D., Zehetner, F., Rampazzo, N., Wimmer, B. and Soja, G. (2016). Long-term effects of biochar on soil physical properties. *Geoderma*, 282, pp. 96–102.

Cao, X., Ma, L., Liang, Y., Gao, B. and Harris, W. (2011). Simultaneous immobilization of lead and atrazine in contaminated soils using dairy-manure biochar. *Environ. Sci. Technol.*, 45(11), pp. 4884–4889.

Cárdenas-Aguiar, E., Gasco, G., Paz-Ferreiro, J. and Méndez, A. (2017). The effect of biochar and compost from urban organic waste on plant biomass and properties of an artificially copper polluted soil. *Int. Biodeterior. Biodegradation*, 124, pp. 223–232.

Chan, K.Y., Van Zwieten, L., Meszaros, I., Downie, A. and Joseph, S. (2008). Using poultry litter biochars as soil amendments. *Aust. J. Soil Res.*, 46, pp. 437–444.

Chen, Y., Xu, J., Lv, Z., Xie, R., Huang, L. and Jiang, J. (2018). Impacts of biochar and oyster shells waste on the immobilization of arsenic in highly contaminated soils. *J. Environ. Manage.*, 217, pp. 646–653.

Choppala, G. K., Bolan, N. S., Megharaj, M., Chen, Z. and Naidu, R. (2012). The influence of biochar and black carbon on reduction and bioavailability of chromate in soils. *J. Environ. Qual.*, 41, pp. 1175–1184.

De Brouwere, K., Smolders, E. and Merckx, R. (2004). Soil properties affecting solid-liquid distribution of As (V) in soils. *Eur. J. Soil Sci.* 55, pp. 165–173.

Debela, F., Thring, R. W. and Arocena, J. M. (2012). Immobilization of heavy metals by co-pyrolysis of contaminated soil with woody biomass. *Water Air Soil Pollut.*, 223(3), pp.1161–1170.

Dhiman, S. S., Selvaraj, C., Li, J., Singh, R., Zhao, X., Kim, D. and Lee, J. K. (2016). Phytoremediation of metal-contaminated soils by the hyperaccumulator canola (Brassica napus L.) and the use of its biomass for ethanol production. *Fuel*, 183, pp. 107–114.

Fellet, G., Marchiol, L., Delle Vedove, G. and Peressotti, A. (2011). Application of biochar on mine tailings: Effects and perspectives for land reclamation. *Chemosphere*, 83, pp. 1262–1297.

Fellet, G., Marmiroli, M. and Marchiol, L. (2014). Elements uptake by metal accumulator species grown on mine tailings amended with three types of biochar. *Sci. Total Environ.*, 468, pp. 598–608.

Gondek, K., Mierzwa-Hersztek, M. and Kopeć, M. (2018). Mobility of heavy metals in sandy soil after application of composts produced from maize straw, sewage sludge and biochar. *J. Environ. Manage.*, 210, pp. 87–95.

Gregory, S. J., Anderson, C. W. N., Camps Arbestain, M. and McManus, M. T. (2014). Response of plant and soil microbes to biochar amendment of an arsenic-contaminated soil. *Agric. Ecosyst. Environ.*, 191, pp. 133–141.

Hartley, W., Dickinson, N. M., Riby, P. and Lepp, N. W. (2009). Arsenic mobility in brownfield soils amended with green waste compost or biochar and planted with Miscanthus. *Environ. Pollut.*, 157, pp. 2654–2662.

Herath, H. M. S. K., Camps-Arbestain, M. and Hedley, M. (2013). Effect of biochar on soil physical properties in two contrasting soils: An Alfisol and an Andisol. *Geoderma*, 209, pp. 188–197.

Houben, D., Evrard, L. and Sonnet, P. (2013a). Mobility, bioavailability and pH-dependent leaching of cadmium, zinc and lead in a contaminated soil amended with biochar. *Chemosphere*, 92(11), pp. 1450–1457.

Houben, D., Evrard, L. and Sonnet, P. (2013b). Beneficial effects of biochar application to contaminated soils on the bioavailability of Cd, Pb and Zn and the biomass production of rapeseed (*Brassica napus* L.). *Biomass Bioenergy*, 57, pp. 196–204.

Hseu, Z. Y., Jien, S. H., Chien, W. H. and Liou, R. C. (2014). Impacts of biochar on physical properties and erosion potential of a mudstone slopeland soil. *The Scientific World Journal*, 2014, pp. 1–10.

Jiang, J., Xu, R., Jiang, T. and Li, Z. (2012). Immobilization of Cu (II), Pb (II) and Cd (II) by the addition of rice straw derived biochar to a simulated polluted Ultisol. *J. Hazard Mater.*, 229–230, pp. 145–150.

Jien, S. H. and Wang, C. S. (2013). Effects of biochar on soil properties and erosion potential in a highly weathered soil. *Catena*, 110, pp. 225–233.

Kambo, H. S. and Dutta, A. (2015). A comparative review of biochar and hydrochar in terms of production, physico-chemical properties and applications. *Renew. Sustain. Energy Rev.*, 45, pp. 359–378.

Karami, N., Clemente, R., Moreno-Jiménez, E., Lepp, N. W. and Beesley, L. (2011). Efficiency of green waste compost and biochar soil amendments for reducing lead and copper mobility and uptake to ryegrass. *J. Hazard. Mater.*, 191(1–3), pp. 41–48.

Karhu, K., Mattila, T., Bergström, I. and Regina, K. (2011). Biochar addition to agricultural soil increased CH4 uptake and water holding capacity — Results from a short-term pilot field study. *Agric. Ecosyst. Environ.*, 140(1–2), pp. 309–313.

Khan, S., Chao, C., Waqas, M., Arp, H. P. H. and Zhu, Y. G. (2013). Sewage sludge biochar influence upon rice (*Oryza sativa* L.) yield, metal bioaccumulation and greenhouse gas emissions from acidic paddy soil. *Environ. Sci. Technol.*, 47(15), pp. 8624–8632.

Kloss, S., Zehetner, F., Oburger, E., Buecker, J., Kitzler, B., Wenzel, W. W. *et al.* (2014). Trace element concentrations in leachates and mustard plant tissue (*Sinapis alba* L.) after biochar application to temperate soils. *Sci. Total Environ.*, 481, pp. 498–508.

Kuppusamy, S., Thavamani, P., Mcgharaj, M., Venkateswarlu, K. and Naidu, R. (2016). Agronomic and remedial benefits and risks of applying biochar to soil: Current knowledge and future research directions. *Environ. Int.*, 87, pp. 1–12.

Laird, D. A., Fleming, P., Davis, D. D., Horton, R., Wang, B. and Karlen, D. L. (2010). Impact of biochar amendments on the quality of a typical Midwestern agricultural soil. *Geoderma*, 158(3–4), pp. 443–449.

Lehmann, J., Pereira da Silva, J., Steiner, C., Nehls, T., Zech, W. and Glaser, B. (2003). Nutrient availability and leaching in an archaeological Anthrosol and a Ferralsol of the Central Amazon basin: Fertilizer, manure and charcoal amendments. *Plant Soil.*, 249, pp. 343–357.

Lehmann, J., Rillig, M. C., Thies, J., Masiello, C. A., Hockaday, W. C. and Crowley, D. (2011). Biochar effects on soil biota — A review. *Soil Biol. Biochem.*, 43(9), pp. 1812–1836.

Li, Z., Jia, M., Christie, P., Ali, S. and Wu, L. (2018). Use of a hyperaccumulator and biochar to remediate an acid soil highly contaminated with trace metals and/or oxytetracycline. *Chemosphere*, 204, pp. 390–397.

Li, X., Wang, X., Chen, Y., Yang, X. and Cui, Z. (2019). Optimization of combined phytoremediation for heavy metal contaminated mine tailings by a field-scale orthogonal experiment. *Ecotoxicol. Environ. Saf.*, 168, pp. 1–8.

Liu, Y., Zhu, J., Ye, C., Zhu, P., Ba, Q., Pang, J. and Shu, L. (2018a). Effects of biochar application on the abundance and community composition of denitrifying bacteria in a reclaimed soil from coal mining subsidence area. *Sci. Total Environ.*, 625, pp. 1218–1224.

Liu, L., Li, W., Song, W. and Guo, M. (2018b). Remediation techniques for heavy metal-contaminated soils: Principles and applicability. *Sci. Total Environ.*, 633, pp. 206–219.

Lomaglio, T., Hattab-Hambli, N., Bret, A., Miard, F., Trupiano, D., Scippa, G. S. *et al.* (2017). Effect of biochar amendments on the mobility and (bio) availability of As, Sb and Pb in a contaminated mine technosol. *J. Geochem. Explor.*, 182, pp. 138–148.

Lu, H., Li, Z., Fu, S., Méndez, A., Gascó, G. and Paz-Ferreiro, J. (2014). Can biochar and phytoextractors be jointly used for cadmium remediation? *PLoS One*, 9, pp. 1–7.

Lu, H., Li, Z., Fu, S., Méndez, A., Gascó, G. and Paz-Ferreiro, J. (2015). Combining phytoextraction and biochar addition improves soil biochemical properties in a soil contaminated with Cd. *Chemosphere*, 119, pp. 209–216.

Lu, H. P., Li, Z. A., Gascó, G., Méndez, A., Shen, Y. and Paz-Ferreiro, J. (2018). Use of magnetic biochars for the immobilization of heavy metals in a multi-contaminated soil. *Sci. Total Environ.*, 622–623, pp. 892–899.

Ma, J. W., Wang, H. and Luo, Q. S. (2007). Movement-adsorption and its mechanism of Cd in soil under combining effects of electrokinetics and a new type of bamboo charcoal. *Huan Jing Ke Xue.*, 28(8), pp. 1829–1834.

Méndez, A., Gómez, A., Paz-Ferreiro, J. and Gascó, G. (2012). Effects of sewage sludge biochar on plant metal availability after application to a Mediterranean soil. *Chemosphere*, 89, pp. 1354–1359.

Moreno-Jiménez, E., Manzano, R., Esteban, E. and Peñalosa, J. M. (2012). The fate of arsenic in soils adjacent to an old-mine site (Bustarviejo, Spain): Mobility and transfer to native flora. *J. Soils Sediments*, 10, pp. 301–312.

Mukherjee, A. and Lal, R. (2013). Biochar impacts on soil physical properties and greenhouse gas emissions. *Agronomy*, 3(2), pp. 313–339.

Namgay, T., Singh, B. and Singh, B. P. (2010). Influence of biochar application to soil on the availability of As, Cd, Cu, Pb, and Zn to maize (*Zea mays* L.). *J. Aust. Soil. Res.*, 48, pp. 638–647.

Novak, J. M., Busscher, W. J., Watts, D. W., Amonette, J. E., Ippolito, J. A., Lima, I. M. *et al.* (2012). Biochars impact on soil-moisture storage in an ultisol and two aridisols. *Soil Sci.*, 177(5), pp. 310–320.

Novak, J. M., Ippolito, J. A., Ducey, T. F., Watts, D. W., Spokas, K. A., Trippe, K. M. *et al.* (2018). Remediation of an acidic mine spoil: Miscanthus biochar and lime amendment affects metal availability, plant growth, and soil enzyme activity. *Chemosphere*, 205, pp. 709–718.

Park, J. H., Choppala, G. K., Bolan, N. S., Chung, J. W. and Chuasavathi, T. (2011). Biochar reduces the bioavailability and phytotoxicity of heavy metals. *Plant Soil.*, 348, pp. 439–451.

Park, J. H., Choppala, G., Lee, S. J., Bolan, N., Chung, J. W. and Edraki, M. (2013). Comparative sorption of Pb and Cd by biochars and its implication for metal immobilization in soils. *Water. Air. Soil Pollut.*, 224, pp. 1–12.

Paz-Ferreiro, J., Gascó, G., Gutiérrez, B. and Méndez, A. (2012). Soil biochemical activities and the geometric mean of enzyme activities after application of sewage sludge and sewage sludge biochar to soil. *Biol. Fertil. Soils.*, 48, pp. 511–517.

Paz-Ferreiro, J., Lu, H., Fu, S., Méndez, A. and Gascó, G. (2014). Use of phytoremediation and biochar to remediate heavy metal polluted soils: A review. *Solid Earth*, 5, pp. 65–75.

Prendergast-Miller, M. T., Duvall, M. and Sohi, S. P. (2014). Biochar–root interactions are mediated by biochar nutrient content and impacts on soil nutrient availability. *Eur. J. Soil Sci.*, 65(1), pp. 173–185.

Puga, A. P., Abreu, C. A., Melo, L. C. A. and Beesley, L. (2015). Biochar application to a contaminated soil reduces the availability and plant uptake of zinc, lead and cadmium. *J. Environ. Manage.*, 159, pp. 86–93.

Puga, A. P., Melo, L. C. A., de Abreu, C. A., Coscione, A. R. and Paz-Ferreiro, J. (2016). Leaching and fractionation of heavy metals in mining soils amended with biochar. *Soil Tillage Res.*, 164, pp. 25–33.

Rees, F., Sterckeman, T. and Morel, J. L. (2016). Root development of non-accumulating and hyperaccumulating plants in metal-contaminated soils amended with biochar. *Chemosphere*, 142, pp. 48–55.

Rodríguez-Vila, A., Covelo, E. F., Forján, R. and Asensio, V. (2014). Phytoremediating a copper mine soil with Brassica juncea L., compost and biochar. *Environ. Sci. Pollut. Res.*, 21(19), pp. 11293–11304.

Rodríguez-Vila, A., Covelo, E. F., Forján, R. and Asensio, V. (2015). Recovering a copper mine soil using organic amendments and phytomanagement with Brassica juncea L. *J. Environ. Manage.*, 147, pp. 73–80.

Rue, M., Rees, F., Simonnot, M. O. and Morel, J. L. (2019). Phytoextraction of Ni from a toxic industrial sludge amended with biochar. *J. Geochem. Explor.*, 196, pp. 173–181.

Sarwar, N., Imran, M., Shaheen, M. R., Ishaque, W., Kamran, M. A., Matloob, A. *et al.* (2017). Phytoremediation strategies for soils contaminated with heavy metals: Modifications and future perspectives. *Chemosphere*, 171, pp. 710–721.

Shrestha, R. K. and Lal, R. (2011). Changes in Physical and Chemical Properties of Soil after Surface Mining and Reclamation. *Geoderma*, 161(3–4), pp. 168–76.

Sitarz-Palczak, E. and Kalembkiewicz, J. (2012). Study of Remediation of Soil Contaminated with Heavy Metals by Coal Fly Ash, *J. Environ. Prot.*, 3, pp. 1373–1383.

Smith, E., Naidu, R. and Alston, A.M. (2002). Chemistry of inorganic arsenic in soils. II. Effect of phosphorous, sodium, and calcium on arsenic sorption. *J. Environ. Qual.*, 31, pp. 557–563.

Stachowicz, M., Hiemstra, T. and van Riemsdijk, W. H. (2008). Multi-competitive interaction of As (III) and As (V) oxyanions with Ca2+, Mg2+, PO4 3–, and CO3 2– ions on goethite. *J. Colloid Interface Sci.*, 320, pp. 400–414.

Sun, Z., Moldrup, P., Elsgaard, L., Arthur, E., Bruun, E. W., Hauggaard-Nielsen, H. and de Jonge, L. W. (2013). Direct and indirect short-term effects of biochar on physical characteristics of an arable sandy loam. *Soil Sci.*, 178(9), pp. 465–473.

Suppadit, T., Kitikoon, V., Phubphol, A. and Neumnoi, P. (2012). Effect of quail litter biochar on productivity of four new physic nut varieties planted in cadmium-contaminated soil. *Chilean J. Agric. Res.*, 72, pp. 125–132.

Uchimiya, M., Lima, I. M., Klasson, K. T. and Wartelle, L. H. (2010). Contaminant immobilization and nutrient release by biochar soil amendment: Roles of natural organic matter. *Chemosphere*, 80(8), pp. 935–940.

Van der Ent, A., Baker, A. J., Reeves, R. D., Chaney, R. L., Anderson, C. W., Meech, J. A. *et al.* (2015). Agromining: Farming for metals in the future? *Environ. Sci. Technol.*, 49(8), pp. 4773–4780.

Wong, M. H. (2003). Ecological restoration of mine degraded soils, with emphasis on metal contaminated soils. *Chemosphere*, 50(6), pp. 775–780.

Xia, H. P. and Cai, X. A. (2002). Ecological restoration technologies for mined lands: A review. *J. Appl. Ecol.*, 13(11), pp. 1471–1477.

Xu, P., Sun, C. X., Ye, X. Z., Xiao, W. D., Zhang, Q. and Wang, Q. (2016). The effect of biochar and crop straws on heavy metal bioavailability and plant accumulation in a Cd and Pb polluted soil. *Ecotoxicol. Environ. Saf.*, 132, pp. 94–100.

Zhang, F. P., Li, C. F., Tong, L. G., Yue, L. X., Li, P., Ciren, Y. J. and Cao, C. G. (2010). Response of microbial characteristics to heavy metal pollution of mining soils in central Tibet, China. *Appl. Soil Ecol.*, 45(3), pp. 144–151.

Zhang, X., Wang, H., He, L., Lu, K., Sarmah, A., Li, J., Bolan, N.S., Pei, J. and Huang, H. (2013). Using biochar for remediation of soils contaminated with heavy metals and organic pollutants. *Environ. Sci. Pollut. Res. Int.*, 20, pp. 8472–8483.

Zhang, R. H., Li, Z. G., Liu, X. D., Wang, B. C., Zhou, G. L., Huang, X. X. *et al.* (2017). Immobilization and bioavailability of heavy metals in greenhouse soils amended with rice straw-derived biochar. *Ecol. Eng.*, 98, pp. 183–188.

Zhou, J. B., Deng, C. J., Chen, J. L. and Zhang, Q. S. (2008). Remediation effects of cotton stalk carbon on cadmium (Cd) contaminated soil. *Ecol. Environ.*, 17, pp. 1857–1860.

Zornoza, R., Acosta, J. A., Faz, A. and Bååth, E. (2016). Microbial growth and community structure in acid mine soils after addition of different amendments for soil reclamation. *Geoderma*, 272, pp. 64–72.

Chapter 6

Biochar-based Carbon Materials for Adsorptive Separation and Applications in Catalysis

Tamara L. Church, Anthony E. Szego and Niklas Hedin*

Department of Materials and Environmental Chemistry, Stockholm University, SE-109 61 Stockholm, Sweden

*niklas.hedin@mmk.su.se

Abstract

Carbon-based sorbents derived from sustainable materials have the potential to replace those derived from fossil-based resources. Biochars have great prospects as sorbents and precursors for sorbents, with targeted uses in both gas cleaning and water treatment. In this chapter, such materials are reviewed and discussed together with related carbon materials and their applications in adsorption processes (e.g. separation of CO_2 from N_2-rich flue gas, biogas upgrading and removal of aqueous pollutants such as dyes, phenols and metal cations) as well as in catalytic or electrocatalytic transformations (e.g. transesterification and hydrogenation reactions, reduction of NOx, ozonation of gaseous ammonia and oxygen reduction and evolution reactions).

6.1 Introduction

Biochars and materials derived from them can be used in adsorptive separations and catalysis, either directly or after being converted to activated carbons (ACs). ACs are used in commercial purification applications, and are produced by activating carbon-rich precursors. Here, we will consider only applications involving materials derived from biochars, but will discuss uses of both as-synthesized and activated biochars, and we therefore briefly introduce the process of activation.

Biochar can be "physically activated" by heating ($\geq 600°C$) with oxidizing gases (e.g. $H_2O(g)$, CO_2, dilute O_2 in N_2) or "chemically activated" by heating with activating agents such as H_3PO_4 or KOH. Though extensively studied in the academic literature, ACs chemically activated with KOH have only been commercialized for high-end applications. Without further insight into the commercial aspects of ACs, it is difficult to assess why this is the case, but based on the assumption that this relates to the corrosive nature of KOH and related safety concerns, other K^+-based activation methods are being studied [Sevilla *et al.*, 2017].

Currently, many commercialized ACs are made by first pyrolyzing biomass (e.g. coconut shell) to form a precursor that is activated, typically with CO_2 or $H_2O(g)$, in a second step. ACs can be prepared from biochars that are made by pyrolysis ("pyrolytic biochars") or by hydrothermal carbonization (HTC, to give "hydrochar"). ACs from pyrolytic biochars have been applied in the separation of CO_2 from flue gas [Yang *et al.*, 2017; Manyà *et al.*, 2018; Li *et al.*, 2015; Deng *et al.*, 2014; Plaza *et al.*, 2014; Alabadi *et al.*, 2015], purification of gas streams from H_2S or NH_3 [Bhandari *et al.*, 2014; Hervy *et al.*, 2018], water purification [Moreno-Virgen *et al.*, 2012], catalysis [Xiong *et al.*, 2017; Cha *et al.*, 2010], and as catalyst supports [Yan *et al.*, 2013; Kim *et al.*, 2016].

In general, hydrochars have advantages over pyrolytic biochars in terms of chemical modification and co-polymerization [Demir-Cakan *et al.*, 2009] as well as thermal balance and carbon-atom efficiency [Hu *et al.*, 2010]. ACs have been synthesized from hydrochars by physical activation in CO_2 [Hao *et al.*, 2013] or $H_2O(g)$ [Azargohar and Dalai, 2008] and by chemical activation with KOH [Falco *et al.*, 2013] or other K^+-containing compounds [Sevilla *et al.*, 2017], H_3PO_4 [Hao *et al.*, 2014], and $LiCl/ZnCl_2$ [Fechler *et al.*, 2013]. Still, more research is needed to judge whether hydrochar-derived ACs can reach the functionalities of ACs derived from pyrolytic biochar, in particular with respect to gas purification.

6.2 Adsorption

6.2.1 *Gas adsorption*

Activated carbons and the related carbon molecular sieves (CMSs) are used commercially to purify and upgrade several gas mixtures. However, most studies of biochar-derived ACs have been focused on the potential separation of CO_2 from N_2-rich flue gas, as adsorption has been suggested as a method to lower the cost of carbon-capture processes [Abanades *et al.*, 2015].

6.2.1.1 *Adsorbing CO_2 from flue gas*

Flue gases from the combustion of fossil fuels consist of a dilute stream of CO_2 in N_2 at a moderately high temperature and close to ambient pressure. Selecting and optimizing ACs for use in adsorption-driven processes requires understanding the thermodynamic state of the flue gas, and the details of the adsorption technology selected. Normally, it is assumed that flue gases cannot be easily compressed or cooled at the scale of power plants. This assumption has been challenged [Liu *et al.*, 2014], and the ability to compress or cool flue gases on a large scale has been opened up for new approaches to adsorbent design. However, under the assumption that these processes cannot be made economical, vacuum swing adsorption (VSA) and possibly temperature swing adsorption (TSA) are the processes available for CO_2 removal with ACs. In a thermally well-integrated power plant, the flue gas temperature is about 50°C and the relative pressure of CO_2 is 0.05–0.15 bar. When comparing ACs for CO_2 capture, it is important to assess the uptake under these conditions.

   CO_2 capture from flue gas requires ACs with high ultramicropore volume, which can be achieved by activation with KOH [Deng *et al.*, 2014] or CO_2 [Hao *et al.*, 2013]. There has been some discussion of whether N-doped carbons could be better CO_2 adsorbents than regular ACs, but recent studies indicate that this is not the case [Sevilla *et al.*, 2013]. The suitability of ACs for VSA-driven CO_2 capture from flue gas can be judged from the ultramicropore volume or the uptake of CO_2 at a low pressure. The uptake of CO_2 at 0.15 bar and 0°C is often reported and can be used for comparison (see selected values in Table 6.1). ACs from pyrolytic biochar take up more CO_2 than those from hydrochar, at least after activation with KOH. The study by Deng *et al.* [2014] and the entries in Table 6.1 indicate that the conditions of activation (time, temperature,

Table 6.1: Uptake of CO_2 on activated carbons prepared from biomass.

Precursor	Agent ($m_{biochar}$:m_{agent})	Atm.[a]	T (°C)	t (h)	CO_2 uptake at 0.15 bar, and 0°C (mmol/g)	Ref.[b]
Hydrochar	KOH (1:3.3), wet impregnation	N_2	700	4	1.8	Hao *et al.* [2017]
Hydrochar	$K_2C_2O_4$ (1:3.6), dry mixing	N_2	800	1	1.4 (25°C)	Sevilla *et al.* [2018]
Hydrochar	KOH (1:2)	N_2	700	1	1.7	Sevilla and Fuertes [2011]
Hydrochar	CO_2	CO_2	800	2	1.8	Hao *et al.* [2013]
Hydrochar	LiCl/$ZnCl_2$	N_2	900	2	1.2	Kumar *et al.* [2017]
Hydrochar	None	N_2	950	2	1.9	Yu *et al.* [2012]
Pyrolytic biochar	KOH (1:2), wet impregnation	N_2	650	1	2.3	Yang *et al.* [2017]
Pyrolytic biochar	KOH (1:1), wet impregnation	N_2	700	1	2.42	Manyà *et al.* [2018]
Pyrolytic biochar	KOH (1:1), wet impregnation	N_2	640	1	2.7	Li *et al.* [2015]
Pyrolytic biochar	KOH (1:2), wet impregnation	N_2	700	1.5	3.3	Bhandari *et al.* [2014]
Biomass	3% O_2 in N_2		650	1.8	1.7	Plaza *et al.* [2014]

Notes: [a] Atm. = Atmosphere; [b] Ref. = References.

amount of KOH) should be optimized, and that KOH addition via wet impregnation seems preferred.

Notably, most studies have been performed with KOH as an activating agent even though its commercial relevance for large-scale applications is challenging. It would therefore be relevant to perform further detailed studies of the use of CO_2 activation or similar for the preparation of microporous ACs from both hydrochars and pyrolytic biochars. Finally, we note that it might be more appropriate with respect to gas separation to report the volumetric uptake level in units of (v/v).

The amount of ultramicropores in an AC is often maximized at a relatively low temperature of activation, whereas micropores and mesopores are induced at higher activation temperatures [Deng *et al.*, 2014; Hao *et al.*, 2017; Sevilla *et al.*, 2018]. The study of Sevilla *et al.* [2018]

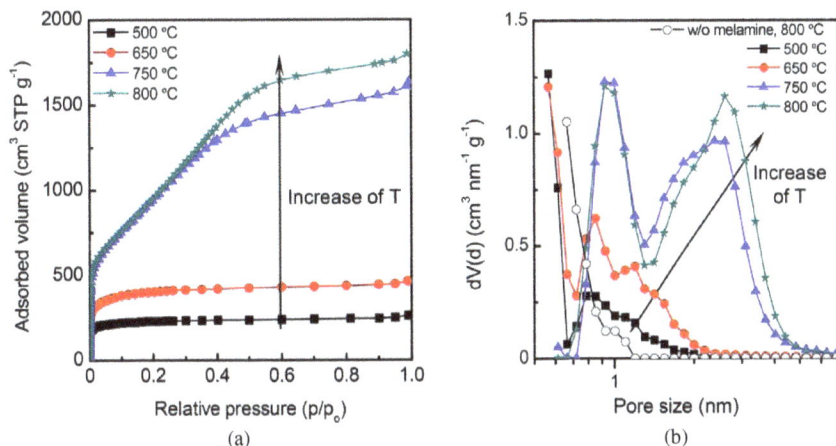

Figure 6.1:   (a) N_2 sorption isotherms and (b) pore size distributions for activated carbons prepared from hydrochars without and with melamine (melamine:glucose ratio of 2:1) under chemical activation with potassium oxalate.

Source: Reproduced from Sevilla *et al.* [2018].

presents an elegant example of the control of pore size via the activation of a hydrochar made from glucose and melamine (Figure 6.1). The very large mass flow of flue gas requires the uptake and release of CO_2 from a sorbent to be very rapid, which could motivate the use of ACs with meso-porosity, and one could argue that regular mesoporosity, which can be induced by hard- [Jun *et al.*, 2000] or soft-templating [Liang and Dai, 2006] would be best. In addition, if flue gas can be compressed or chilled at scale, mesoporous ACs with high total pore volume would be preferred. Structured monoliths could provide better process efficiency than gran-ules [Akhtar *et al.*, 2014].

Much less effort has been devoted to designing ACs or other porous carbons derived from biochars for TSA processes. Zhao *et al.* [2010] made some initial studies of amine-modified hydrochars, and recently, Chatterjee *et al.* [2018] supported tetraethylenepentamine on pyrolytic biochars to produce a material with a high CO_2 adsorption capacity (2.79 mmol/g at 70°C). Supported, tethered or polymerized primary amines are relevant for TSA processes as they react readily with CO_2 at the tempera-ture of flue gas and release it at a higher temperature. Amines have been supported on carbons for this application [Adelodun *et al.*, 2015]. However, certain cost estimates indicate that TSA-driven CO_2 capture is significantly more expensive than its VSA-driven analogue [Abanades *et al.*, 2015].

6.2.1.2 *Other relevant gas treatment and storage processes*

Pressure swing adsorption (PSA) and VSA are today used for the separation of N_2 from air and CO_2 from biogas. For these processes, CMSs are used as their narrow pores can enhance the N_2–over–O_2 or the CO_2–over–CH_4 selectivity by differentially affecting the respective diffusion rates of the gases. Some ACs derived from hydrochars have narrow pore size distributions [Borrero-López *et al.*, 2017], but are not synthesized by the chemical vapor deposition approach used for CMSs, and it would be relevant to study whether such ACs could be used in molecular sieving.

H_2S can be removed from various gas streams with ACs, and is important in biogas upgrading. ACs derived from pyrolytic biochars have been studied for this application. H_2S adsorbs strongly to ACs and for this application it is important that the micropore volume of the AC is high. The adsorption of H_2S is difficult to study for safety reasons, but can be measured on the AC in a column [Hervy *et al.*, 2018]. The adsorption of NH_3 on ACs is also relevant but technically challenging to measure, and has been studied to some degree on ACs derived from pyrolytic biochars [Bhandari *et al.*, 2014].

The adsorptive storage of CH_4 is an alternative to compression and liquefaction and is relevant for onboard and intermediate storage of CH_4. ACs have been studied for this application in some detail [Matranga *et al.*, 1992]. Here too, the volumetric capacity (v/v) might be more relevant than the gravimetric, and monoliths can have advantages over pellets. A uniform carbon structure and high volumetric adsorption capacity are advantageous [Stadie *et al.*, 2013]. Important work has been performed on deriving ACs from hydrochar for this application [Falco *et al.*, 2013; Mestre *et al.*, 2014; Li *et al.*, 2016], but more research into CH_4 uptake on zeolite-templated carbons is needed. The requirements for the onboard storage of CH_4 set by agencies such as the ARPA-E (315% (v/v)) are quite challenging.

6.2.2 *Adsorption from aqueous solution*

Studies of biochar and ACs from biomass sources to adsorb aqueous solutes, in particular pollutant species (e.g. dyes, phenols, metal cations, oxyanions), are copious. The cost and environmental impact of a biochar are lowest when it is produced from a waste product and used after minimal transport. Thus, a wide variety of biochars from precursors abundant in a given geographical area have been tested as adsorbents for pollutants relevant to the area, and it is not necessarily useful or practical to directly compare these studies, as test conditions vary. We give an overview of the methods used to evaluate the removal of contaminants (sorbates) from water by biochars, and discuss the chemical interactions involved. Recent examples have been chosen to illustrate the molecular interactions and measurements involved in adsorption on biochars rather than according to the risk posed by the pollutant; priorities for pollutant removal are available elsewhere [WHO, 2017]. Recent review articles offer tabulated adsorption capacities of biochars for aqueous species [Sophia and Lima, 2018; Afroze and Sen, 2018; Wei *et al.*, 2018; Li *et al.*, 2017; Ncibi *et al.*, 2017; Yin *et al.*, 2017; Cha *et al.*, 2016; De Gisi *et al.*, 2016; Inyang and Dickenson, 2015; Tan *et al.*, 2015; Mohan *et al.*, 2014].

6.2.2.1 *Evaluation of biochars as sorbents*

The performance of a sorbent is generally evaluated in terms of sorption capacity, i.e. the amount of a sorbate that it can remove from a solution, and sorption kinetics, i.e. the rate at which it removes the sorbate.

Sorption capacity

The sorption capacity of a char, i.e. the amount of a particular sorbate that it can adsorb at equilibrium, is most simply measured as an adsorption percentage, where the amount adsorbed from the solution by a biochar is compared to the amount initially present. This measure is valid only under the specific conditions measured (initial sorbate concentration $[Q]_0$, sorbent loading, T, pH, etc.), but it is useful in evaluating the effectiveness of a sorbent dose in a specific application, such as the adsorption percentage of Pb^{2+} on biochar-derived ACs from wastewater from battery manufacturing [Thitame and Shukla, 2017; Abdel-Galil *et al.*, 2016].

More information about the interaction of a dissolved species with a sorbent can be obtained using an isotherm. Often, biochars are exposed to solutions of different $[Q]_0$ at a single temperature. For each measurement, the amount of the species adsorbed per unit mass of sorbent at equilibrium, q_{eq}, is compared to the concentration $[Q]_{eq}$ remaining in the solution, and the relationship is modeled mathematically. The two most commonly invoked isotherms, the Langmuir [1918] and Freundlich [1907] isotherms, are covered here; a comprehensive review of isotherm types is available [Foo and Hameed, 2010].

The Langmuir isotherm (Eq. (6.1)) is by far the most commonly used model to describe the capacity of a biochar sorbent to adsorb a species from water. This model assumes that a sorbent surface is adorned with a fixed number of identical, independent adsorption sites. The adsorption of a species to these sites is described by an equilibrium constant b, and the number of available sites in the monolayer is described by the maximum sorption capacity q_{max} (per unit sorbent mass).

$$q_{eq} = \frac{q_{max}b[Q]_{eq}}{1 - b[Q]_{eq}} \tag{6.1}$$

The Langmuir isotherm is often linearized, but with the result that correlated variables are plotted; therefore, data should be analyzed using the nonlinear form of the isotherm (Eq. (6.1)) with an appropriate error model [Foo and Hameed, 2010]. A sufficient range of solution concentrations should be tested, and if possible, maximum sorption capacity should be measured rather than predicted from the isotherm [Harter, 1984]. Equation (6.1) provides a suitable description of experimental data in many systems; although materials as heterogeneous as biochars would not be expected to bear uniform adsorption sites, representing the surface this way often yields a useful two-parameter description of sorption that has a physical basis. Values of q_{max} have been published for many sorbent/sorbate pairs. When the assumptions underlying the Langmuir model hold true, b is a thermodynamic equilibrium constant related to ΔG for sorption and, if measured at multiple temperatures, can be used to calculate ΔH and ΔS using the Van't Hoff equation. Preferably, a global numeric analysis of the temperature- and concentration-dependent isotherms uniquely describing the thermodynamics of the adsorbent–solute system should be made. Although Eq. (6.1) produces very good correlation coefficients for many sorption reactions and provides a convenient quantity to compare among sorbents, it does not always describe measured data well; as Langmuir stated, "… the phenomena of adsorption by porous bodies are inherently very complex and that we should not expect them to be represented by a simple formula".

The Langmuir model has been extended by for example Brunauer *et al.* [1938] (BET model).

The Freundlich isotherm (Eq. (6.2)) also relates the equilibrium values q_{eq} and $[Q]_{eq}$. Though developed empirically, the Freundlich isotherm has been reconstructed on a thermodynamically sound base by assuming adsorption sites having an exponentially decaying distribution of the heat of sorption [Jaroniec, 1975]. A three-parameter isotherm that takes the form in Eq. (6.2) at low concentrations of adsorbate, but reaches a maximum adsorption capacity at higher concentrations, has been developed [Sips, 1948].

$$q_{eq} = K_F [Q]_{eq}^{1/n} \qquad (6.2)$$

Concentration-dependent isotherms are useful in describing the capacity of a sorbent for a pollutant and for comparing sorbents, but do not capture the temperature dependence, and will not necessarily hold when the adsorption conditions (e.g. pH) are changed. Information regarding how changes in conditions impact sorption on a biochar can be obtained by combining isotherm data with data from other sources. For example, the adsorption of Ni^{2+} and Zn^{2+} on biochar was studied using isotherms obtained at multiple pH values, spectroscopic and calorimetric information, and data regarding the speciation of the ions in water [Alam *et al.*, 2018]. Such detailed investigation is not necessary when screening sorbents, but the predictive model obtained is valuable for promising materials.

Related to the capacity of a sorbent is the question of regeneration and reuse. It is not necessarily true that biochar sorbents should be reused; as they are preferably derived from abundant, inexpensive waste biomass, and regeneration processes require the use of additional materials, it may be more economical to use a biochar sorbent only once [Tan *et al.*, 2015]. However, regeneration and reuse could conceivably be worthwhile if a biochar has been chemically modified to produce a more effective sorbent. Economic and environmental evaluations of this issue are needed, but the technical feasibility of reusing biochar sorbents has been demonstrated. For example, ethylenediaminetetraacetic (EDTA) acid and its disodium salt have been used to leach Pb^{2+} and Cd^{2+} from functionalized biochar-derived ACs [Ribeiro *et al.*, 2015; Wang *et al.*, 2015], enabling their reuse, though with aqueous solutions of EDTA and heavy metal ions as byproducts.

Sorption kinetics

There exist various mathematical models to describe sorption kinetics, and we show the three most common here. Restraint in the mechanistic interpretation of these models is encouraged.

Lagergren [1898] found a linear relationship between the instantaneous rate at which organic acids adsorbed onto acid-washed animal charcoal and the difference between the final and instantaneous amounts sorbed, i.e. to the sorption sites available at the measured instant. This situation is called "pseudo-first order kinetics" or sometimes "Lagergren kinetics", and for times between $t = 0$ and $t = t$, takes the integrated form in Eq. (6.3).

$$q_t = q_\infty (1 - e^{-kt}) \qquad (6.3)$$

where q_i is the amount adsorbed per sorbent mass at time i; k is a rate constant.

This rate equation has been used to describe sorption on many sorbents; however, in many cases, a pseudo-second-order model, in which the instantaneous rate of adsorption is related to the square of the available sorption sites (integrated form in Eq. (6.4)), better describes the experimental data [Ho and McKay, 1999]. The order that best describes sorption data can depend on initial concentration of sorbate though, with a second-order model providing a better description for low concentrations of sorbate [Azizian, 2004]. The integrated pseudo-second-order equation can be rearranged to the linear form in Eq. (6.5), which allows data to be fit to the model by plotting t/q_t vs. t. However, this plot involves correlated variables, and is therefore expected to give higher correlation coefficients.

$$\frac{1}{(q_\infty - q_t)} = \frac{1}{q_\infty} + kt \qquad (6.4)$$

$$\frac{t}{q_t} = \frac{1}{kq_\infty^2} + \frac{t}{q_\infty} \qquad (6.5)$$

In cases where the diffusion of a sorbate within a sorbent particle determines the sorption rate, various kinds of intraparticle diffusion models are used. These models can include cases where the rate of sorption is set by a surface layer of the particle, or when the diffusion is controlled by the intraparticle diffusion rates. In the latter case, and assuming iso-thermal sorption by a spherical particle with radius a and a sorbate characterized by the diffusion coefficient D, the time-dependent uptake is described by Eq. (6.6) [Crank, 1975]. For ACs, the intraparticle diffusion rate could be time-dependent, as it depends implicitly on concentration and the pores in ACs are at a similar scale as the adsorbate.

$$\frac{q_t}{q_\infty} = 1 - \frac{6}{\pi^2} \sum_{n=1}^{\infty} \frac{1}{n^2} e^{\frac{-Dtn^2\pi^2}{a^2}} \tag{6.6}$$

In real applications, the extended incubation of wastewater with a biochar sorbent is not practical, and dynamic tests are needed. Thus, water containing a pollutant can be passed through a column packed with biochar, and the time at which the pollutant is eluted ("breakthrough") can be studied. Biochar that was generated from corn cobs and then modified via sequential treatment with HNO_3 and NaOH was packed into columns and tested as a sorbent for NH_4^+(aq) [Nguyen *et al.*, 2017]. In another report, pyrolytic biochar from date seeds was used to remove Ni^{2+} and Cu^{2+} ions from a solution [Mahdi *et al.*, 2018]. Powdered biochar is not suitable for real column systems, so biochar–biopolymer composite beads have also been tested as sorbents in fixed-bed columns. For example, chars from rice straw [Jang *et al.*, 2018] and microalgae [Jung *et al.*, 2017] were combined with sodium alginate and washed with $CaCl_2$(aq) to form biochar–calcium alginate sorbents that removed Sr^{2+} or phosphate ($H_nPO_4^{(3-n)-}$) from water. Using the biochar–alginate composite that had been used to adsorb phosphate as a soil additive improved the growth of lettuce (*Lactuca sativa*) [Jung *et al.*, 2017].

6.2.2.2 *Relevant chemical interactions and reaction conditions*

Manifold chemical interactions mediate sorption to biochars; the most relevant ones will vary depending upon the properties of the biochar and the sorbate, and on the sorption conditions. The interactions that are most important to the sorption of metal ions [Li *et al.*, 2017], metals and organics [Tan *et al.*, 2015; Abbas *et al.*, 2018], organic and microbial components [Inyang and Dickenson, 2015; Pignatello *et al.*, 2017], and ionic and ionizable organics [Kah *et al.*, 2017] have been reviewed. Here, we present a brief overview of the interactions that mediate sorption to biochars.

Note that biochars, being heterogeneous in nature and bearing diverse functional groups, can interact with a single sorbate in multiple ways. Further, many types of interactions cannot be observed diagnostically, but rather are inferred based upon correlations between biochar or sorbate properties and sorption capacity; however, one material property or reaction condition can impact multiple types of chemical and physical interactions.

pH, surface charge and speciation

Prior to discussing how biochars interact with molecules in aqueous solution, it is useful to mention pH, which impacts the nature of the sorbent and sorbate. As discussed in earlier chapters of this book, the biochar surface bears many chemical functionalities; in an aqueous medium, the form taken by ionizable functional groups on a biochar surface depends on the pH of the medium. Carboxylic acid groups are neutral when the pH of the medium is below their pK_a, but exist as negatively charged carboxylates at higher pH. Phenols can be deprotonated, though at higher pH than carboxylic acids. Amine groups can be protonated if the solution pH is below their pK_b. Thus the surface charge and chemistry of a biochar in H_2O varies significantly with pH, and it is relevant to determine the pH_{PZC} (PZC = point of zero charge) at which a material surface is not charged. In solutions having pH $<$ pH_{PZC}, a material will have a positive surface charge; for pH $>$ pH_{PZC}, the surface will be negatively charged.

Dissolved sorbate species, especially metal ions, can also take different forms in solution depending on pH. For example, lead exists in aqueous solution as Pb^{2+} at pH $<$ 6, but can precipitate as $Pb(OH)_2$ at higher pH values (Figure 6.2(a)). Similarly, many organic contaminants contain ionizable functional groups (e.g. carboxylates, amines, phenols), and a single organic molecule can contain several such groups [Kah *et al.*, 2017]. As pH affects the functional groups and charge on the biochar surface as well as the identity and charge of sorbate species in the solution, its influence on adsorption reactions can be complex.

The speciation of some metal species is also affected by redox reactions. Metals and metalloids in highly oxidized forms, such as Cr^{VI} (Figure 6.2(b)) and As^V, can be reduced by electron-rich groups that are part of a solid biochar, or that are released into the solution from a biochar. Notably, solution pH will have different impacts on the speciation, charge and solubility of the oxidized and reduced metal ions (Figure 6.2). This can profoundly affect whether a species is adsorbed onto biochar. For example, a biochar that was positively charged at pH = 2 removed

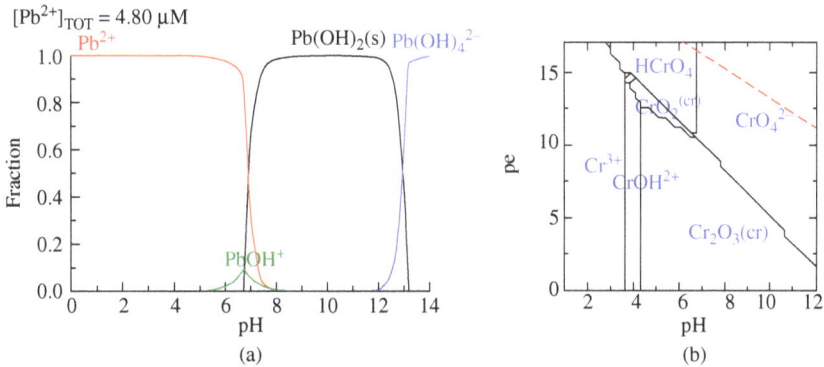

Figure 6.2: Speciation of (a) aqueous Pb(II) as a function of pH, and (b) aqueous chromium ions as a function of pH and reduction potential. Speciation is shown for 100 × the allowable concentrations in drinking water [WHO, 2017]; thus [Pb] = 4.8 × 10^{-6} M and [Cr] = 9.62 × 10^{-5} M.

Source: Diagrams produced using MEDUSA software [Puigdomenech, 2010].

Cr^{VI} (present as negatively charged HCrO$_4^-$ at low pH; Figure 6.2(b)), but very little Cr^{III} (positively charged at pH 2) from the solution under those conditions [Choudhary *et al.*, 2017]. Both the biochar itself, and organics that dissolved from it, reduced Cr^{VI} to Cr^{III} in the solution, so reduction competed with adsorption. Adsorbed Cr was present primarily as Cr^{III}, so surface-bound Cr could be reduced to Cr^{III} and retained on the surface; it was the speciation of Cr in the solution that was decisive for sorption.

Non-specific interactions

Hydrophobic interactions can be important to the sorption of neutral organic molecules on biochars that have been produced at moderate or high temperature, and are thus more carbonized. These interactions should be most relevant for the sorption of molecules with non-polar domains, and their participation in a sorption process can therefore be indicated by a positive relationship between the octanol–water partition coefficient (K_{O-W}) and the sorption capacity for a solute. For example, the biocides chlorpyrifos and diuron (Figure 6.3, log K_{O-W} = 4.7 and 2.87) were adsorbed to a much greater degree than glyphosate (log K_{O-W} = −3.2) on a biochar produced from the pyrolysis of wood at 405 ± 25°C [Cederlund *et al.*, 2016]. Alternatively, the Kamlet–Taft polarity/polarizability

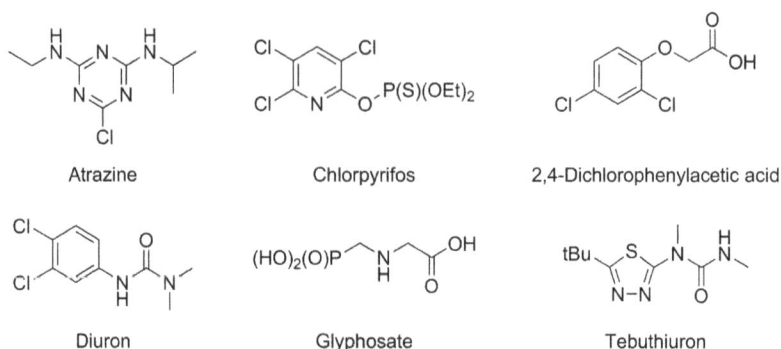

Figure 6.3:   Organic compounds discussed in this section.

parameter for a solute can indicate its hydrophobicity. The values of this and other parameters for a set of 25 aromatic compounds were compared to the sorption affinity of six biochars for these compounds, and the resulting correlations used to estimate the contribution of various interactions to the sorption of each compound to each biochar [Yang *et al.*, 2018]. In all cases, hydrophobic interactions were the main contributor, which is reasonable given that the biochars examined were generated at 700°C.

Although weak on a molar basis, *van der Waals interactions* are additive, and are important in sorption. Micropores (d < 2 nm) have high surface area where such interactions can accumulate, and are therefore favorable sorption sites for non- or weakly polar molecules that are able to enter the pores. This concentration of non-specific interactions leads to the "pore-filling" mechanism of sorption. Further, confinement in very small pores can impact the properties of a sorbate. In some cases, confined substances behave as solids at temperatures where the unconfined substance behaves as a liquid; this is particularly observed in porous solids with attractive surfaces (e.g. carbon, mica) [Alba-Simionesco *et al.*, 2006]. The contribution of pore filling to sorption is often assessed by examining the ability of a biochar to adsorb chemically similar substrates that vary in size, or how a solute is adsorbed by biochars with varying pore sizes.

For biochars produced at relatively low temperatures (≤400°C), non-specific interactions can lead to *absorptive partitioning* into the non-carbonized organic domains of a char, similar to the dissolution of compounds in the organic matter of soils [Chiou *et al.*, 1979]. Such partitioning can be slow, but increases linearly with the concentration of the sorbate, and is not subject to competition from other sorbates, which could be an important advantage in complex solutions. The insensitivity of such partitioning to competition has been used to distinguish the contributions of ab- and adsorption mechanisms for the uptake of substituted aromatics on biochars [Chiou *et al.*, 2015]. Absorption was more important for the uptake of sorbates by biochars produced at 250°C, whereas adsorption was more important on biochars produced at 400°C or 700°C. For ionizable organics, the importance of absorption depends on pH, as neutral and ionic forms of a compound have different solubilities [Kah *et al.*, 2017].

Ions are also subject to non-specific interactions, and the charged surface of a biochar will repel sorbates with like charge and attract oppositely charged species, setting up an electric double layer even at low concentrations of a salt. *Electrostatic interactions* depend strongly on pH because of its impact on the charge and electrostatic potential of the biochar surface and of the sorbate. The ionic strength of a solution will also impact the importance of electrostatic interactions for sorption, as a solution with higher ionic strength can shield charges.

Specific non-covalent interactions

π–π *interactions* can be imagined as electrostatic interactions between the π cloud of an aromatic system with the positive electrostatic potential of the edge of another π-system. The notion of π–π interactions is under discussion [Zhao and Zhang, 2017; Wheeler and Bloom, 2014], and a complete understanding of their character is lacking. Nevertheless, π–π interactions are often invoked to explain the sorption of aqueous species, particularly aromatics, onto biochar. For example, the importance of π–π interactions for the sorption of triazine herbicides and related compounds was demonstrated by comparing the adsorption of the compounds on biochar and on graphite, which is composed of extended sheets of aromatic carbon [Xiao and Pignatello, 2015]. To support the idea that π–π interactions are involved in sorption to graphite, the electrostatic potential of each sorbate molecule at the midpoint of its aromatic ring was correlated to the adsorption coefficient of the sorbate on graphite (normalized to K_{O-W} for the sorbate in order to control for hydrophobic interactions). The strongest effect was observed for sorbates having heteroatoms in the aromatic ring.

Cations interact strongly with negatively charged π clouds via *cation–π interactions* [Dougherty, 2013], and the interaction strength, at least as measured in the gas phase, is enough to be significantly important to the sorption of cations on biochar. The pH of an aqueous solution can impact whether cation–π interactions are important to sorption from that solution, in that it determines whether the sorbate of interest is present as a cation, and whether the overall surface has a relevant electrostatic potential. Moreover, the ionic strength of a water sample can affect the degree to which cation–π interactions are important in sorption on a biochar by introducing competition from cations other than the sorbate of interest.

Specific covalent interactions

The large number of –OH groups on biochar make H-bonds highly probable [Kah *et al.*, 2017] but, in an aqueous medium, the high concentration of water molecules gives them a competitive advantage for the available sites [Pignatello *et al.*, 2017]. However, H-bonds between moieties with similar proton affinities are strong and have partial covalent character [Gilli *et al.*, 2009], and those that fulfill this criterion and involve one or more charged moieties are especially strong. The strength of this type of hydrogen bond, called a *charge-assisted H-bond*, can overcome electrostatic repulsion between like charges, and thus favor the sorption of negatively charged species to a negatively charged biochar surface. The H-bonding groups on a biochar surface are generally carboxylates ($-CO_2^-$) or phenolates ($-O^-$), so hydrogen bonds to these groups are strongest for sorbates with similar pK_a values, i.e. other carboxylic acids or phenols. For example, the sorption of 2,4-dichlorophenoxyacetate (conjugate base of the biocide 2,4-dichlorophenoxyacetic acid, Figure 6.3) on a negatively charged maple-wood-derived biochar, normalized to the micropore surface area of the biochar, increased considerably upon the partial oxidation of the biochar surface [Xiao and Pignatello, 2016]. The same was not true for neutral atrazine or tebuthiuron (Figure 6.3). Further, the sorption of 2,4-dichlorophenoxyacetate was lowered more by competition from added 4-tolylcarboxylate than by competition from the neutral 4-tolyl acetate, despite that the latter had a greater affinity for the biochar.

Carboxylic (as $-CO_2H$ or $-CO_2^-$) and phenolic (as –OH or $-O^-$) groups on the biochar surface can act as ligands for metal ions, and thus cause sorption by *complexation*. Complexation can sometimes be observed spectroscopically. Alternatively, the functional groups on a biochar can be transformed into non- or weakly complexing moieties via organic reactions, and the impact of this change on sorption can be assessed. In a recent example, a biochar produced from the pyrolysis of bagasse at 450°C was used to adsorb Hg^{2+} from the solution [Xu *et al.*, 2016]. Modified biochars in which the $-CO_2H$ or $-OH$ functionalities had been "blocked" adsorbed noticeably less Hg^{2+} from the solution, especially at low [Hg^{2+}]. This information, in combination with spectroscopic data, implicated the formation of covalent complexes between Hg^{2+} from solution and $-CO_2H$ or $-OH$ on the biochar in Hg removal. Complexation to these groups can be important in the sorption of many metal ions to biochars [Li *et al.*, 2017]; in the case of Hg, coordination to S can also be important [Liu *et al.*, 2016].

The complementary case of coordination is also observed; organic compounds can be removed from the aqueous solution when they form complexes with metal ions present in a biochar. For example, ash species, and specifically those containing Fe and Al, were crucial to the sorption of acid-extractable organics (mainly naphthenic acids) from oil sands process water onto biochars derived from straws, grass and wood [Bhuiyan *et al.*, 2017].

Specific ionic interactions

The *precipitation of insoluble salts* is important in the sorption of ions like Pb^{2+} and Cd^{2+}, which are primarily cationic at pH < 7 (see Figure 6.2(a)), by biochars generated from manure, which can have high contents of anions (CO_3^{2-} and PO_4^{3-}) that form insoluble salts with these metals at circumneutral pH [Li *et al.*, 2017]. These microcrystalline precipitates are easily detected using powder X-ray diffraction. For example, biochars produced from the pyrolysis of chicken manure were used to remove Cd^{2+} from aqueous solution [Huang *et al.*, 2018]. The formation of Cd salts was the major contributor to sorption on the chars produced at 700°C, which contained the most PO_4^{3-} and CO_3^{2-}. In contrast, precipitation played a minor role in Cd^{2+} sorption on a plant-based biochar [Zhang *et al.*, 2015]. Solution pH affects salt solubility, and therefore sorption via salt formation; >85% of the Cd^{2+} adsorbed by the pyrolyzed-manure biochar (above) dissolved upon exposure to a solution with pH = 1 [Huang *et al.*, 2018].

Finally, the adsorption of aqueous species to biochars is sometimes described as occurring via *ion exchange*. Fundamentally, ion exchange must be mediated by one of the interactions discussed above (usually complexation or electrostatic interactions), but the term highlights the fact that ions can be released from biochar during sorption, and this must be considered when evaluating the appropriateness of an adsorbent for water purification. For example, for every charge adsorbed as Cd^{2+} onto biochar derived from dried invasive water hyacinth (*Eichhornia crassipes*), nearly 0.8 equivalents of charge were released, primarily as Ca^{2+} and Mg^{2+} but also Na^+ and K^+ [Zhang *et al.*, 2015]. These ions may have been displaced from coordinating oxygen sites. Similarly, biochar that was produced via the pyrolysis of municipal sludge at high temperature released Ca^{2+} and Mg^{2+} when adsorbing cationic Cr^{III} from acidic solution, but released much less Ca^{2+} and Mg^{2+} when adsorbing Cr^{VI}, which exists primarily as oxyanions under these conditions and was poorly adsorbed [Chen *et al.*, 2015].

Compared to oxygen functionalities that support exchangeable cations, biochars contain fewer covalently bound positively charged moieties with exchangeable anions, and anion exchange is therefore less commonly implicated in sorption on biochars. At lower pH, exchangeable anions can be associated with protonated pyridines and perhaps protons bound to the biochar surface via cation–π interactions; at higher pH, they are likely associated with oxonium ions in the biochar matrix [Lawrinenko and Laird, 2015]. Biochar composites with a higher fraction of exchangeable anions have been designed; for example, a composite produced by depositing layered double hydroxides onto biochar derived from bamboo straw removed phosphate from aqueous solution, but lost Cl, presumably as Cl^- [Wan *et al.*, 2017]. Ion exchange is subject to competition from ions other than the target sorbate, so ionic strength can affect the role of ion exchange in adsorption of pollutants.

6.3 Catalysis

The most common use of biochars in catalysis is as supports for metal or metal oxide active sites, but biochar surfaces have oxygen functionalities such as furanic, phenolic or carboxylic oxygen, and these alone can have catalytic activity. Metal or metal oxide particles can be added to a biochar by modifying the conditions of its synthesis or via post-synthesis treatment, and the biochars can be removed by combustion.

Recently, biochar-based materials have been studied as catalysts in refinery processes (e.g. syngas cleaning and conversion), biodiesel production and air-pollution control. It has been suggested that surface functionalities, for example $-SO_3H$ groups acting as Brønsted acid sites, and dispersed metal nanoparticles, are key to the catalytic activity of these materials. Other properties such as particle strength, hydrophobicity, surface area, porosity and mineral content also emerge as important factors for some applications. This section collects uses of biochar-based heterogeneous catalysts in biorefinery (chemical synthesis and biofuel production) and non-biorefinery (pollution control and other non-biorefinery) processes.

6.3.1 *Biorefinery*

6.3.1.1 *Chemical synthesis*

Lignocellulosic biomass is a renewable raw material useful for the synthesis of various platform chemicals, biofuels and materials. It contains carbohydrate polymers (cellulose and hemicellulose) and aromatic polymers (lignin) interacting via intra- and intermolecular hydrogen bonds. Increasing the commercial value of these biopolymers usually involves extensive breakdown and transformation reactions, among which hydrolysis, isomerization, dehydration and rehydration can be facile over catalysts tethered, impregnated, adsorbed on or complexed with biochars or biochar-derived supports. Most of the reactions involved in the breakdown and transformations depend on the Brønsted and Lewis acidity of the catalysts. This acidity can be controlled by adding sulfonated groups, generally by treating the porous carbon in concentrated sulfuric acid at high temperatures, to yield sulfonic acid functionalized solid acid catalysts. For example, Li *et al.* [2013] used sulfonated corn-stover-derived biochar to convert lignocellulosic biomass to glucose (8–10% yield, 19–22% conversion with respect to the corresponding polysaccharide) and xylose (23–41%, 68–81% conversion). The results were comparable to those obtained in the hydrolysis of model compounds over the same catalyst: cellulose gave 3% glucose (24% conversion) and xylan produced 40% xylose (100% conversion). Thus, in this case, the impurities of the complex biomass substrate did not play a major role in the catalytic reactions. The catalysts could be recovered.

Sulfonated biochar has been used to derive platform molecules such as monosaccharides or 5-(hydroxylmethyl)furfural (HMF) from biomass by hydrolysis and dehydration reactions. Hydrochars were prepared from glucose or sucrose and sulfonated with concentrated sulfuric acid [Liu *et al.*, 2013]. Alternative methods of sulfonation such as the direct HTC with sulfonic precursors (mainly hydroxyethylsulfonic acid) have been shown [Nata *et al.*, 2017]. The catalysts prepared using this method presented very high stability and reusability, which are important for future applications.

Functionalized biochars also facilitate the isomerization of glucose to fructose and/or the dehydration of glucose to HMF. Biochar-based sulfonic acids modified by ionic liquid achieved an HMF yield of 28% from cellulose with a moderate selectivity of 62% within 3 h at 80°C in water [Zhang *et al.*, 2017]. The recyclabilities of biochar-based catalysts are important. The activity of a sulfonated biochar for hemicellulose hydrolysis decreased after a first run, and the catalyst was deactivated by the fourth run, possibly because of acid site leaching and mass loss, as reported by Ormsby *et al.* [2012]. However, supported ruthenium particles on mesoporous carbon [Komanoya *et al.*, 2011] did not deactivate after five uses in cellulose hydrolysis, indicating that metal impregnation could be a viable route to biochar-derived hydrolysis catalysts.

6.3.1.2 *Biofuel production*

Esterification/transesterification

Biochar-based catalysts have been applied for biodiesel production. The free fatty acids obtained from vegetable oil and animal fats undergo esterification with low-molecular-weight alcohols to give esters as biofuels. The most used biochar-based materials for esterification and transesterification reactions are the sulfonated hydrochars. Pileidis *et al.* [2014] prepared hydrochars (HTC conditions 230°C, 24 h), converted them into solid acid catalysts, and studied them for the esterification of levulinic acid. Hydrochars were prepared from glucose, cellulose and rice straw, and were sulfonated to produce catalysts with 5–6% sulphur. These were used to catalyze esterification at 60°C and after 3 h almost full conversion and a 97% selectivity toward the ester was achieved with the glucose-derived catalyst.

Similar catalysts have been used for the esterification of glycerol [De la Calle *et al.*, 2015] and oleic acid [Laohapornchaiphan *et al.*, 2017]. In those cases, a glucose-derived hydrochar was sulfonated with a potential inclusion of additional oxygenated groups. The catalytic performance of sulfonated hydrochar in the esterification of glycerol with acetic, butyric and caprylic acids was compared to that of commercial sulfonated resins. Turnover numbers for glycerol and acetic acid were similar for the commercial and hydrochar solid catalysts. The catalysts could be reused after treating with acid to cleave the esterified sulfonic groups that may have formed during the reaction.

In addition to sulfonated biochar, CaO is a powerful heterogeneous alkaline catalyst for biodiesel production [Sirisomboonchai *et al.*, 2015; Lee *et al.*, 2015] so $CaCO_3$-rich biochar (e.g, palm kernel shell, waste egg shell, waste crab shell) has been used as a precursor for transesterification catalysts [Chen *et al.*, 2017].

Hydrogenation

A set of important reactions used to increase the value of biomass-derived syngas are the hydrogenation reactions. Syngas obtained from thermo-chemical conversion of biomass contains CO, H_2, CO_2 and volatile hydro-carbons, which can be upgraded into liquid fuels through e.g. the Fischer–Tropsch synthesis (FTS) and methanation. Iron-nanoparticle-based catalysts encapsulated on biochar-based carbon supports promoted the FTS, achieving 95% CO conversion and 68% selectivity to liquid hydrocarbons [Yan *et al.*, 2013]. Ruthenium particles loaded on activated biochar [Wang *et al.*, 2014] were an effective catalyst for methanation, probably because of the large surface area achievable for these carbon-based supports.

Another use of biochar-based catalysts in the production of biofuels is in hydrogen production through the reforming of biogas [Dufour *et al.*, 2008], though little information on this use of biochar is available. For more extensive information regarding the uses of biochar in biorefinery-related processes, please refer to a review published by Xiong *et al.* [2017].

6.3.2 *Non-biorefinery*

6.3.2.1 *Pollution control*

The main applications of biochar-based catalysts in pollution control are the catalytic reduction of NO_x and the ozonation of gaseous ammonia. NO_x emissions can be efficiently controlled by selectively catalytically reducing them to give N_2. KOH-activated biochars derived from rice straw and sewage sludge removed 86 and 46% of NO_x, respectively [Cha *et al.*, 2010].

The two catalysts had similar surface area and pore volumes, so these properties could not explain the difference in activity. Rather, the minerals and elemental content of each carbon source could affect their catalytic activity. In a review by Cha *et al.* [2016], the addition of metal oxides to biochar-based materials was highlighted as it may significantly increase their catalytic activity for NO_x reduction.

An effective way to remove ammonia is through ozonation (Eq. (6.7)). An activated biochar derived from the pyrolysis of pelletized peanut hulls under steam catalyzed this oxidation of ammonia, but only at high concentrations of ozone [Kastner *et al.*, 2009].

$$2\,NH_3 + 7/3\,O_3 \longrightarrow 2\,NO_2 + 3\,H_2O \qquad (6.7)$$

An emerging application of biochar-based materials in pollution control is as supports for semiconducting nanoparticles that catalyze the photocatalytic degradation of pollutants; this topic was recently reviewed [Mian and Liu, 2018]. In addition to the high surface area and tunable surface chemistry that makes biochars and biochar-derived materials appropriate catalyst supports, their ability to accept and store electrons is advantageous for photocatalytic reactions because it can improve electron–hole separation. Additionally, the adsorbent characteristics of the biochar can concentrate the pollutant as it adsorbs on the char surface [Kim and Kan, 2016].

6.3.2.2 *Other non-biorefinery*

Biochars have been studied as catalysts and catalyst supports for other chemical reactions as well. Biochars, and in particular hydrochars, have polar surfaces and much lower surface areas than ACs, but upon reducing the oxygen content and increasing that of carbon, the properties of hydrochars can become closer to those of a polar AC [Roldán *et al.*, 2015]. By tuning these parameters (surface area and polarity) the deposition of metal precursors can be enhanced. In this respect, palladium nanoparticles supported onto biochar were employed for the Suzuki–Miyaura coupling [Dong *et al.*, 2017]. The catalyst demonstrated high catalytic activity for the reaction of many aryl halides with boronic acids, and could be recycled up to five times through simple centrifugation. In liquid-state reactions, leaching of supported metal is usually a problem, but the properties of biochar favor the redeposition of leached particles during cool-down of the reaction.

Intricate supports can be designed by combining HTC with a porous polymer template, as reported by Cheng *et al.* [2014]. In this case, a polymer template was introduced during the HTC process to give a char that was then treated at 700°C in a reducing atmosphere. This thermal treatment increased the surface area significantly. The biochar was then loaded with Au nanoparticles to produce a highly active catalyst for the reduction of 4-nitrophenol to 4-aminophenol with $NaBH_4$.

A niche application for biochar that has been studied recently is electrocatalysis. Electrocatalysis plays a crucial role in many energy storage and conversion technologies such as the oxygen reduction reaction (ORR) at the cathode of metal–air batteries or fuel cells, the oxygen evolution reaction (OER) and hydrogen evolution reactions (HER) occurring at either electrode of water electrolyzers and CO_2 reduction in liquid fuel conversion devices. These electrocatalytic reactions generally display slow kinetics and as such require catalysts that can improve and give future perspective to these technologies [Tang *et al.*, 2017].

Among the metal-free ORR electrocatalysts, significant research has been focused on nitrogen-doped carbon materials [Alatalo *et al.*, 2016; Lu *et al.*, 2017; Preuss *et al.*, 2017; Yu *et al.*, 2016; Qiao *et al.*, 2016]. Active sites for the ORR are generated inside the graphitic matrix of carbon materials due to the polarization of the C–N bonds (whose polarization strength depends on the type of nitrogen contained), which activates the neighbouring carbon atoms. At the same time, the delocalization of electrons within the graphitic π-system translates into an increase in the conductivity of the material [Ismagilov *et al.*, 2009].

Doping with heteroatoms other than N (e.g. B, S and P) has also been tested [Preuss *et al.*, 2017]. In one study of carbogels derived from glucose and ovalbumin, the synergistic effect of B and N was thought to augment the electron transfer numbers and lower hydrogen peroxide yields when compared to those observed in purely N-doped systems, whereas the presence of S decreased the surface area and N content of the materials, lowering the ORR performance. Heteroatom-doped systems are not only used for ORR, they have also been tested for the HER. 2D crystalline carbons were obtained from hydrochars prepared from saccharides with guanine, which played an important role in producing the 2D-morphology of the resultant carbon materials [Huang *et al.*, 2017]. The porous N-doped carbons were highly active for ORR, and showed efficiency for HER.

In the past, the development of electrochemical water splitting has been held back by the slow kinetics of the OER. Non-noble metal alternatives for OER are often based on transition metal oxides, whereas carbon-based materials have generally been underexplored because of relatively poor performance. However, one approach used peroxovanadium complexes to create oxygen-containing functional groups on the surface of an AC cloth, resulting in a higher specific surface area and faster electron transfer rate compared to a pristine carbon cloth [Huang *et al.*, 2017].

For more information on the applicability of biochar-based materials in non-biorefinery-related processes, please refer to reviews published by Lee *et al.* [2017] and Qian *et al.* [2015].

References

Abanades, J. C., Arias, B., Lyngfelt, A., Mattisson, T., Wiley, D. E., Li, H., Ho, M. T., Mangano, E. and Brandani, S. (2015). Emerging CO_2 capture systems, *Int. J. Greenhouse Gas Control*, 40, pp. 126–166.

Abbas, Z., Ali, S., Rizwan, M., Zaheer, I. E., Malik, A., Riaz, M. A., Shahid, M. R., Zia ur Rehman, M. and Al-Wabel, M. (2018). A critical review of mechanisms involved in the adsorption of organic and inorganic contaminants through biochar. *Arabian J. Geosci.*, 11, p. 448.

Abdel-Galil, E. A., Rizk, H. E. and Mostafa, A. Z. (2016). Isotherm, kinetic, and thermodynamic studies for sorption of Cu(II) and Pb(II) by activated carbon prepared from Leucaena plant wastes. *Part. Sci. Technol.*, 34, pp. 540–551.

Adelodun, A. A., Kim, K.-H., Ngila, J. C. and Szulejko, J. (2015). A review on the effect of amination pretreatment for the selective separation of CO_2. *Appl. Energy*, 158, pp. 631–642.

Afroze, S. and Sen, T. K. (2018). A review on heavy metal ions and dye adsorption from water by agricultural solid waste adsorbents. *Water, Air, Soil Pollut.*, 229, pp. 1–50.

Akhtar, F., Andersson, L., Ogunwumi, S., Hedin, N. and Bergström, L. (2014). Structuring adsorbents and catalysts by processing of porous powders. *J. Eur. Ceram. Soc.*, 34, pp. 1643–1666.

Alabadi, A., Razzaque, S., Yang, Y., Chen, S. and Tan, B. (2015). Highly porous activated carbon materials from carbonized biomass with high CO_2 capturing capacity. *Chem. Eng. J.*, 281, pp. 606–612.

Alam, M. S., Gorman-Lewis, D., Chen, N., Flynn, S. L., Ok, Y. S., Konhauser, K. O. and Alessi, D. S. (2018). Thermodynamic analysis of nickel(II) and zinc(II) adsorption to biochar. *Environ. Sci. Technol.*, 52, pp. 6246–6255.

Alatalo, S.-M., Qiu, K., Preuss, K., Marinovic, A., Sevilla, M., Sillanpää, M., Guo, X. and Titirici, M.-M. (2016). Soy protein directed hydrothermal synthesis of porous carbon aerogels for electrocatalytic oxygen reduction. *Carbon*, 96, pp. 622–630.

Alba-Simionesco, C., Coasne, B., Dosseh, G., Dudziak, G., Gubbins, K. E., Radhakrishnan, R. and Sliwinska-Bartkowiak, M. (2006). Effects of confinement on freezing and melting. *J. Phys.: Condens. Matter*, 18, p. R15.

Azargohar, R. and Dalai, A. K. (2008). Steam and KOH activation of biochar: Experimental and modeling studies. *Microporous Mesoporous Mater.*, 110, pp. 413–421.

Azizian, S. (2004). Kinetic models of sorption: A theoretical analysis. *J. Colloid Interface Sci.*, 276, pp. 47–52.

Bhandari, P. N., Kumar, A. and Huhnke, R. L. (2014). Simultaneous removal of toluene (model tar), NH_3, and H_2S, from biomass-generated producer gas using biochar-based and mixed-metal oxide catalysts. *Energy Fuels*, 28, pp. 1918–1925.

Bhuiyan, T. I., Tak, J. K., Sessarego, S., Harfield, D. and Hill, J. M. (2017). Adsorption of acid-extractable organics from oil sands process-affected water onto biomass-based biochar: Metal content matters. *Chemosphere*, 168, pp. 1337–1344.

Borrero-López, A. M., Fierro, V., Jeder, A., Ouederni, A., Masson, E. and Celzard, A. (2017). High added-value products from the hydrothermal carbonisation of olive stones. *Environ. Sci. Pollut. Res.*, 24, pp. 9859–9869.

Brunauer, S., Emmett, P. H. and Teller, E. (1938). Adsorption of gases in multimolecular layers. *J. Am. Chem. Soc.*, 60, pp. 309–319.

Cederlund, H., Börjesson, E., Lundberg, D. and Stenström, J. (2016). Adsorption of pesticides with different chemical properties to a wood biochar treated with heat and iron. *Water, Air, & Soil Pollution*, 227, p. 203.

Cha, J. S., Choi, J.-C., Ko, J. H., Park, Y.-K., Park, S. H., Jeong, K.-E., Kim, S.-S. and Jeon, J.-K. (2010). The low-temperature SCR of NO over rice straw and sewage sludge derived char. *Chem. Eng. J.*, 156, pp. 321–327.

Cha, J. S., Park, S. H., Jung, S.-C., Ryu, C., Jeon, J.-K., Shin, M.-C. and Park, Y.-K. (2016). Production and utilization of biochar: A review. *J. Ind. Eng. Chem.*, 40, pp. 1–15.

Chatterjee, R., Sajjadi, B., Mattern, D. L., Chen, W.-Y., Zubatiuk, T., Leszczynska, D., Leszczynski, J., Egiebor, N. O. and Hammer, N. (2018). Ultrasound cavitation intensified amine functionalization: A feasible strategy for enhancing CO_2 capture capacity of biochar. *Fuel*, 225, pp. 287–298.

Chen, S. S., Maneerung, T., Tsang, D. C. W., Ok, Y. S. and Wang, C.-H. (2017). Valorization of biomass to hydroxymethylfurfural, levulinic acid, and fatty acid methyl ester by heterogeneous catalysts. *Chem. Eng. J.*, 328, pp. 246–273.

Chen, T., Zhou, Z., Xu, S., Wang, H. and Lu, W. (2015). Adsorption behavior comparison of trivalent and hexavalent chromium on biochar derived from municipal sludge. *Bioresour. Technol.*, 190, pp. 388–394.

Cheng, J., Wang, Y., Teng, C., Shang, Y., Ren, L. and Jiang, B. (2014). Preparation and characterization of monodisperse, micrometer-sized, hierarchically porous carbon spheres as catalyst support. *Chem. Eng. J.*, 242, pp. 285–293.

Chiou, C. T., Peters, L. J. and Freed, V. H. (1979). A physical concept of soil-water equilibria for nonionic organic compounds. *Science*, 206, pp. 831–832.

Chiou, C. T., Cheng, J., Hung, W.-N., Chen, B. and Lin, T.-F. (2015). Resolution of adsorption and partition components of organic compounds on black carbons. *Environ. Sci. Technol.*, 49, pp. 9116–9123.

Choudhary, B., Paul, D., Singh, A. and Gupta, T. (2017). Removal of hexavalent chromium upon interaction with biochar under acidic conditions: mechanistic insights and application. *Environ. Sci. Pollut. Res.*, 24, pp. 16786–16797.

Crank, J. (1975). *The Mathematics of Diffusion.* 2nd ed., (Oxford Univ. Press: Oxford, UK).

De Gisi, S., Lofrano, G., Grassi, M. and Notarnicola, M. (2016). Characteristics and adsorption capacities of low-cost sorbents for wastewater treatment: A review. *Sustainable Mater. Technol.*, 9, pp. 10–40.

De la Calle, C., Fraile, J. M., García-Bordejé, E., Pires, E. and Roldán, L. (2015). Biobased catalyst in biorefinery processes: sulphonated hydrothermal carbon for glycerol esterification. *Catal. Sci. Technol.*, 5, pp. 2897–2903.

Demir-Cakan, R., Baccile, N., Antonietti, M. and Titirici, M.-M. (2009). Carboxylate-rich carbonaceous materials via one-step hydrothermal carbonization of glucose in the presence of acrylic acid. *Chem. Mater.*, 21, pp. 484–490.

Deng, S., Wei, H., Chen, T., Wang, B., Huang, J. and Yu, G. (2014). Superior CO_2 adsorption on pine nut shell-derived activated carbons and the effective micropores at different temperatures. *Chem. Eng. J.*, 253, pp. 46–54.

Dong, W., Cheng, S., Feng, C., Shang, N., Gao, S., Wang, C. and Wang, Z. (2017). Carbon nanospheres with well-controlled nano-morphologies as support for palladium-catalyzed Suzuki coupling reaction. *Appl. Organometal. Chem.*, 31, p. e3741.

Dougherty, D. A. (2013). The Cation–π Interaction. *Acc. Chem. Res.*, 46, pp. 885–893.

Dufour, A., Celzard, A., Fierro, V., Martin, E., Broust, F. and Zoulalian, A. (2008). Catalytic decomposition of methane over a wood char concurrently activated by a pyrolysis gas. *Appl. Catal. A*, 346, pp. 164–173.

Falco, C., Marco-Lozar, J. P., Salinas-Torres, D., Morallón, E., Cazorla-Amorós, D., Titirici, M. M. and Lozano-Castelló, D. (2013). Tailoring the porosity of chemically activated hydrothermal carbons: Influence of the precursor and hydrothermal carbonization temperature. *Carbon*, 62, pp. 346–355.

Fechler, N., Fellinger, T.-P. and Antonietti, M. (2013). "Salt Templating": A simple and sustainable pathway toward highly porous functional carbons from ionic liquids. *Adv. Mater.*, 25, pp. 75–79.

Freundlich, H. (1907). Über die Adsorption in Lösungen. *Zeit. Phys. Chem.*, 57U, pp. 385–470.

Foo, K. Y. and Hameed, B. H. (2010). Insights into the modeling of adsorption isotherm systems. *Chem. Eng. J.*, 156, pp. 2–10.

Gilli, P., Pretto, L., Bertolasi, V. and Gilli, G. (2009). Predicting hydrogen-bond strengths from acid–base molecular properties. The pKa slide rule: Toward the solution of a long-lasting problem. *Acc. Chem. Res.*, 42, pp. 33–44.

Hao, W., Björkman, E., Lilliestråle, M. and Hedin, N. (2013). Activated carbons prepared from hydrothermally carbonized waste biomass used as adsorbents for CO_2. *Appl. Energy*, 112, pp. 526–532.

Hao, W., Björkman, E., Lilliestråle, M. and Hedin, N. (2014). Activated carbons for water treatment prepared by phosphoric acid activation of hydrothermally treated beer waste. *Ind. Eng. Chem. Res.*, 53, pp. 15389–15397.

Hao, W., Björnerbäck, F., Trushkina, Y., Bengoechea, M. O., Salazar-Alvarez, G., Barth, T. and Hedin, N. (2017). High-Performance magnetic activated carbon from solid waste from lignin conversion processes. 1. Their use as adsorbents for CO_2. *ACS Sustainable Chem. Eng.*, 5, pp. 3087–3095.

Harter, R. D. (1984). Curve-fit Errors in Langmuir Adsorption Maxima. *Soil Sci. Soc. Am. J.*, 48, pp. 749–752.

Hervy, M., Pham Minh, D., Gerente, C., Weiss-Hortala, E., Nzihou, A., Villot, A. and Le Coq, L. (2018). H_2S removal from syngas using wastes pyrolysis chars. *Chem. Eng. J.*, 334, pp. 2179–2189.

Ho, Y. S. and McKay, G. (1999). Pseudo-second order model for sorption processes. *Process Biochem.*, 34, pp. 451–465.

Hu, B., Wang, K., Wu, L., Yu, S., Antonietti, M. and Titirici, M.-M. (2010). Engineering Carbon Materials from the Hydrothermal Carbonization Process of Biomass. *Adv. Mater.*, 22, pp. 813–828.

Huang, B., Liu, Y. and Xie, Z. (2017). Biomass derived 2D carbons via a hydrothermal carbonization method as efficient bifunctional ORR/HER electrocatalysts, *J. Mater. Chem. A*, 5, pp. 23481–23488.

Huang, D., Li, S., Zhang, X., Luo, Y., Xiao, J. and Chen, H. (2018). A novel method to significantly boost the electrocatalytic activity of carbon cloth for oxygen evolution reaction, *Carbon*, 129, pp. 468–475.

Huang, F., Gao, L.-Y., Deng, J.-H., Chen, S.-H. and Cai, K.-Z. (2018). Quantitative contribution of Cd^{2+} adsorption mechanisms by chicken-manure-derived biochars, *Environ. Sci. Pollut. Res.*, 25, pp. 28322–28334.

Inyang, M. and Dickenson, E. (2015). The potential role of biochar in the removal of organic and microbial contaminants from potable and reuse water: A review, *Chemosphere*, 134, pp. 232–240.

Ismagilov, Z. R., Shalagina, A. E., Podyacheva, O. Y., Ischenko, A. V., Kibis, L. S., Boronin, A. I., Chesalov, Y. A., Kochubey, D. I., Romanenko, A. I., Anikeeva, O. B., Buryakov, T. I. and Tkachev, E. N. (2009). Structure and electrical conductivity of nitrogen-doped carbon nanofibers. *Carbon*, 47, pp. 1922–1929.

Jang, J., Miran, W., Divine, S. D., Nawaz, M., Shahzad, A., Woo, S. H. and Lee, D. S. (2018). Rice straw-based biochar beads for the removal of radioactive strontium from aqueous solution. *Sci. Total Environ.*, 615, pp. 698–707.

Jaroniec, M. (1975). Adsorption on heterogeneous surfaces: The exponential equation for the overall adsorption isotherm. *Surf. Sci.*, 50, pp. 553–564.

Jun, S., Joo, S. H., Ryoo, R., Kruk, M., Jaroniec, M., Liu, Z., Ohsuna, T. and Terasaki, O. (2000). Synthesis of new, nanoporous carbon with hexagonally ordered mesostructure. *J. Am. Chem. Soc.*, 122, pp. 10712–10713.

Jung, K.-W., Jeong, T.-U., Choi, B. H., Kang, H.-J. and Ahn, K.-H. (2017). Phosphate adsorption from aqueous solution by Laminaria japonica-derived biochar-calcium alginate beads in a fixed-bed column: Experiments and prediction of breakthrough curves. *Environ. Prog. Sustainable Energy*, 36, pp. 1365–1373.

Kah, M., Sigmund, G., Xiao, F. and Hofmann, T. (2017). Sorption of ionizable and ionic organic compounds to biochar, activated carbon and other carbonaceous materials. *Water Res.*, 124, pp. 673–692.

Kastner, J. R., Miller, J., Kolar, P. and Das, K. C. (2009). Catalytic ozonation of ammonia using biomass char and wood fly ash. *Chemosphere*, 75, pp. 739–744.

Kim, J. R. and Kan, E. (2016). Heterogeneous photocatalytic degradation of sulfamethoxazole in water using a biochar-supported TiO_2 photocatalyst. *J. Environ. Manage.*, 180, pp. 94–101.

Komanoya, T., Kobayashi, H., Hara, K., Chun, W.-J. and Fukuoka, A. (2011). Catalysis and characterization of carbon-supported ruthenium for cellulose hydrolysis. *Appl. Catal., A*, 407, pp. 188–194.

Kumar, K. V., Gadipelli, S., Preuss, K., Porwal, H., Zhao, T., Guo, Z. X. and Titirici, M.-M. (2017). Salt templating with pore padding: Hierarchical pore tailoring towards functionalised porous carbons. *ChemSusChem*, 10, pp. 199–209.

Lagergren, S. (1898). Zur Theorie der sogenannten Adsorption gelöster Stoffe. *Kongl. Svenska Vet.-Akad. Handl., Bihang*, 24, pp. 1–39.

Laohapornchaiphan, J., Smith, C. B. and Smith, S. M. (2017). One-step preparation of carbon-based solid acid catalyst from water hyacinth leaves for esterification of oleic acid and dehydration of xylose. *Chem. Asian J.*, 12, pp. 3178–3186.

Langmuir, I. (1918). The adsorption of gases on plane surfaces of glass, mica and platinum. *J. Am. Chem. Soc.*, 40, pp. 1361–1403.

Lawrinenko, M. and Laird, D. A. (2015). Anion exchange capacity of biochar. *Green Chem.*, 17, pp. 4628–4636.

Lee, J., Kim, K.-H. and Kwon, E. E. (2017). Biochar as a catalyst. *Renewable Sustainable Energy Rev.*, 77, pp. 70–79.

Lee, S. L., Wong, Y. C., Tan, Y. P. and Yew, S. Y. (2015). Transesterification of palm oil to biodiesel by using waste obtuse horn shell-derived CaO catalyst. *Energy Convers. Manage.*, 93, pp. 282–288.

Li, D., Ma, T., Zhang, R., Tian, Y. and Qiao, Y. (2015). Preparation of porous carbons with high low-pressure CO_2 uptake by KOH activation of rice husk char. *Fuel*, 139, pp. 68–70.

Li, H., Dong, X., da Silva, E. B., de Oliveira, L. M., Chen, Y. and Ma, L. Q. (2017). Mechanisms of metal sorption by biochars: Biochar characteristics and modifications. *Chemosphere*, 178, pp. 466–478.

Li, S., Gu, Z., Bjornson, B. E. and Muthukumarappan, A. (2013). Biochar based solid acid catalyst hydrolyze biomass. *J. Environ. Chem. Eng.*, 1, pp. 1174–1181.

Li, Y., Li, D., Rao, Y., Zhao, X. and Wu, M. (2016). Superior CO_2, CH_4, and H_2 uptakes over ultrahigh-surface-area carbon spheres prepared from sustainable biomass-derived char by CO_2 activation. *Carbon*, 105, pp. 454–462.

Liang, C. and Dai, S. (2006). Synthesis of mesoporous carbon materials via enhanced hydrogen-bonding interaction. *J. Am. Chem. Soc.*, 128, pp. 5316–5317.

Liu, L., Sanders, E. S., Kulkarni, S. S., Hasse, D. J. and Koros, W. J. (2014). Sub-ambient temperature flue gas carbon dioxide capture via Matrimid (R) hollow fiber membranes. *J. Membr. Sci.*, 465, pp. 49–55.

Liu, M., Jia, S., Gong, Y., Song, C. and Guo, X. (2013). Effective Hydrolysis of Cellulose into Glucose over Sulfonated Sugar-Derived Carbon in an Ionic Liquid. *Ind. Eng. Chem. Res.*, 52, pp. 8167–8173.

Liu, P., Ptacek, C. J., Blowes, D. W. and Landis, R. C. (2016). Mechanisms of mercury removal by biochars produced from different feedstocks determined using X-ray absorption spectroscopy, *J. Hazard. Mater.*, 308, pp. 233–242.

Lu, Y., Wang, L., Preuß, K., Qiao, M., Titirici, M.-M., Varcoe, J. and Cai, Q. (2017). Halloysite-derived nitrogen doped carbon electrocatalysts for anion exchange membrane fuel cells. *J. Power Sources*, 372, pp. 82–90.

Mahdi, Z., Yu, Q. J. and El Hanandeh, A. (2018). Investigation of the kinetics and mechanisms of nickel and copper ions adsorption from aqueous solutions by date seed derived biochar. *J. Environ. Chem. Eng.*, 6, pp. 1171–1181.

Manyà, J. J., González, B., Azuara, M. and Arner, G. (2018). Ultra-microporous adsorbents prepared from vine shoots-derived biochar with high CO_2 uptake and CO_2/N_2 selectivity. *Chem. Eng. J.*, 345, pp. 631–639.

Matranga, K. R., Myers, A. L. and Glandt, E. D. (1992). Storage of Natural-Gas by Adsorption on Activated Carbon. *Chem. Eng. Sci.*, 47, pp. 1569–1579.

Mestre, A. S., Freire, C., Pires, J., Carvalho, A. P. and Pinto, M. L. (2014). High performance microspherical activated carbons for methane storage and landfill gas or biogas upgrade. *J. Mater. Chem. A*, 2, pp. 15337–15344.

Mian, M. M. and Liu, G. (2018). Recent progress in biochar-supported photocatalysts: synthesis, role of biochar, and applications. *RSC Adv.*, 8, pp. 14237–14248.

Mohan, D., Sarswat, A., Ok, Y. S. and Pittman Jr., C. U. (2014). Organic and inorganic contaminants removal from water with biochar, a renewable, low cost and sustainable adsorbent: A critical review. *Bioresour. Technol.*, 160, pp. 191–202.

Moreno-Virgen, M. R., Tovar-Gómez, R., Mendoza-Castillo, D. I. and Bonilla-Petriciolet, A. (2012). *Lignocellulosic Precursors Used in the Synthesis of Activated Carbon*, In: Hernández Montoya, V. and Bonilla-Petriciolet, A. (eds.), "Applications of Activated Carbons Obtained from Lignocellulosic Materials for the Wastewater Treatment" (IntechOpen, Rijeka) pp. Ch. 4.

Nata, I. F., Putra, M. D., Irawan, C. and Lee, C.-K. (2017). Catalytic performance of sulfonated carbon-based solid acid catalyst on esterification of waste cooking oil for biodiesel production. *J. Environ. Chem. Eng.*, 5, pp. 2171–2175.

Ncibi, M. C., Mahjoub, B., Mahjoub, O. and Sillanpää, M. (2017). Remediation of emerging pollutants in contaminated wastewater and aquatic environments: biomass-based technologies. *Clean: Soil, Air, Water*, 45, p. 1700101.

Nguyen, L. H., Vu, T. M., Le, T. T., Trinh, V. T., Tran, T. P. and Van, H. T. (2017). Ammonium removal from aqueous solutions by fixed-bed column using corncob-based modified biochar. *Environ. Technol.*, pp. 1–10.

Ormsby, R., Kastner, J. R. and Miller, J. (2012). Hemicellulose hydrolysis using solid acid catalysts generated from biochar. *Catal. Today*, 190, pp. 89–97.

Pignatello, J. J., Mitch, W. A. and Xu, W. (2017). Activity and Reactivity of Pyrogenic Carbonaceous Matter toward Organic Compounds. *Environ. Sci. Technol.*, 51, pp. 8893–8908.

Pileidis, F. D., Tabassum, M., Coutts, S. and Titirici, M.-M. (2014). Esterification of levulinic acid into ethyl levulinate catalysed by sulfonated hydrothermal carbons. *Chin. J. Catal.*, 35, pp. 929–936.

Plaza, M. G., González, A. S., Pis, J. J., Rubiera, F. and Pevida, C. (2014). Production of microporous biochars by single-step oxidation: Effect of activation conditions on CO_2 capture. *Appl. Energy*, 114, pp. 551–562.

Preuss, K., Tănase, L. C., Teodorescu, C. M., Abrahams, I. and Titirici, M.-M. (2017). Sustainable metal-free carbogels as oxygen reduction electrocatalysts. *J. Mater. Chem. A*, 5, pp. 16336–16343.

Puigdomenech, I. (2010). *MEDUSA: Make Equilibrium Diagrams Using Sophisticated Algorithms*, https://www.kth.se/che/medusa/.

Qian, K., Kumar, A., Zhang, H., Bellmer, D. and Huhnke, R. (2015). Recent advances in utilization of biochar, *Renewable Sustainable Energy Rev.*, 42, pp. 1055–1064.

Qiao, M., Tang, C., He, G., Qiu, K., Binions, R., Parkin, I. P., Zhang, Q., Guo, Z. and Titirici, M. M. (2016). Graphene/nitrogen-doped porous carbon sandwiches for the metal-free oxygen reduction reaction: Conductivity versus active sites. *J. Mater. Chem. A*, 4, pp. 12658–12666.

Ribeiro, R. F. L., Soares, V. C., Costa, L. M. and Nascentes, C. C. (2015). Production of activated carbon from biodiesel solid residues: An alternative for hazardous metal sorption from aqueous solution. *J. Environ. Manage.*, 162, pp. 123–131.

Roldán, L., Pires, E., Fraile, J. M. and García-Bordejé, E. (2015). Impact of sulfonated hydrothermal carbon texture and surface chemistry on its catalytic performance in esterification reaction. *Catal. Today*, 249, pp. 153–160.

Sevilla, M. and Fuertes, A. B. (2011). Sustainable porous carbons with a superior performance for CO_2 capture. *Energy Environ. Sci.*, 4, pp. 1765–1771.

Sevilla, M., Parra, J. B. and Fuertes, A. B. (2013). Assessment of the role of micropore size and N-Doping in CO_2 capture by porous carbons. *ACS Appl. Mater. Interfaces*, 5, pp. 6360–6368.

Sevilla, M., Ferrero, G. A. and Fuertes, A. B. (2017). Beyond KOH activation for the synthesis of superactivated carbons from hydrochar. *Carbon*, 114, pp. 50–58.

Sevilla, M., Al-Jumialy, A. S. M., Fuertes, A. B. and Mokaya, R. (2018). Optimization of the pore structure of biomass-based carbons in relation to their use for CO_2 capture under low- and high-pressure regimes. *ACS Appl. Mater. Interfaces*, 10, pp. 1623–1633.

Sips, R. (1948). On the structure of a catalyst surface. *J. Chem. Phys.*, 16, pp. 490–495.

Sirisomboonchai, S., Abuduwayiti, M., Guan, G., Samart, C., Abliz, S., Hao, X., Kusakabe, K. and Abudula, A. (2015). Biodiesel production from waste cooking oil using calcined scallop shell as catalyst. *Energy Convers. Manage.*, 95, pp. 242–247.

Sophia A., C. and Lima, E. C. (2018). Removal of emerging contaminants from the environment by adsorption. *Ecotoxicol. Environ. Saf.*, 150, pp. 1–17.

Stadie, N. P., Murialdo, M., Ahn, C. C. and Fultz, B. (2013). Anomalous isosteric enthalpy of adsorption of methane on zeolite-templated carbon. *J. Am. Chem. Soc.*, 135, pp. 990–993.

Tan, X., Liu, Y., Zeng, G., Wang, X., Hu, X., Gu, Y. and Yang, Z. (2015). Application of biochar for the removal of pollutants from aqueous solutions. *Chemosphere*, 125, pp. 70–85.

Tang, C., Titirici, M.-M. and Zhang, Q. (2017). A review of nanocarbons in energy electrocatalysis: Multifunctional substrates and highly active sites. *J. Energy Chem.*, 26, pp. 1077–1093.

Thitame, P. V. and Shukla, S. R. (2017). Removal of lead (II) from synthetic solution and industry wastewater using almond shell activated carbon. *Environ. Prog. Sustainable Energy*, 36, pp. 1628–1633.

Wan, S., Wang, S., Li, Y. and Gao, B. (2017). Functionalizing biochar with Mg–Al and Mg–Fe layered double hydroxides for removal of phosphate from aqueous solutions. *J. Ind. Eng. Chem.*, 47, pp. 246–253.

Wang, S., Wang, H., Yin, Q., Zhu, L. and Yin, S. (2014). Methanation of bio-syngas over a biochar supported catalyst. *New J. Chem.*, 38, pp. 4471–4477.

Wang, S.-Y., Tang, Y.-K., Chen, C., Wu, J.-T., Huang, Z., Mo, Y.-Y., Zhang, K.-X. and Chen, J.-B. (2015). Regeneration of magnetic biochar derived from eucalyptus leaf residue for lead(II) removal. *Bioresour. Technol.*, 186, pp. 360–364.

Wei, D., Li, B., Huang, H., Luo, L., Zhang, J., Yang, Y., Guo, J., Tang, L., Zeng, G. and Zhou, Y. (2018). Biochar-based functional materials in the purification of agricultural wastewater: Fabrication, application and future research needs. *Chemosphere*, 197, pp. 165–180.

Wheeler, S. E. and Bloom, J. W. G. (2014). Toward a More Complete Understanding of Noncovalent Interactions Involving Aromatic Rings. *J. Phys. Chem. A*, 118, pp. 6133–6147.

World Health Organization, (2017) *Guidelines for Drinking-Water Quality*, 4th edition Ed. "Chemical aspects" (World Health Organization, Geneva) pp. 155–201.

Xiao, F. and Pignatello, J. J. (2015). Interactions of triazine herbicides with biochar: Steric and electronic effects. *Water Res.*, 80, pp. 179–188.

Xiao, F. and Pignatello, J. J. (2016). Effects of post-pyrolysis air oxidation of biomass chars on adsorption of neutral and ionizable compounds. *Environ. Sci. Technol.*, 50, pp. 6276–6283.

Xiong, X., Yu, I. K. M., Cao, L., Tsang, D. C. W., Zhang, S. and Ok, Y. S. (2017). A review of biochar-based catalysts for chemical synthesis, biofuel production, and pollution control. *Bioresour. Technol.*, 246, pp. 254–270.

Xu, X., Schierz, A., Xu, N. and Cao, X. (2016). Comparison of the characteristics and mechanisms of Hg(II) sorption by biochars and activated carbon. *J. Colloid Interface Sci.*, 463, pp. 55–60.

Yan, Q., Wan, C., Liu, J., Gao, J., Yu, F., Zhang, J. and Cai, Z. (2013). Iron nanoparticles in situ encapsulated in biochar-based carbon as an effective catalyst for the conversion of biomass-derived syngas to liquid hydrocarbons. *Green Chem.*, 15, pp. 1631–1640.

Yang, K., Jiang, Y., Yang, J. and Lin, D. (2018). Correlations and adsorption mechanisms of aromatic compounds on biochars produced from various biomass at 700°C. *Environ. Pollut.*, 233, pp. 64–70.

Yang, J., Yue, L., Hu, X., Wang, L., Zhao, Y., Lin, Y., Sun, Y., DaCosta, H. and Guo, L. (2017). Efficient CO_2 capture by porous carbons derived from coconut shell. *Energy Fuels*, 31, pp. 4287–4293.

Yin, Q., Zhang, B., Wang, R. and Zhao, Z. (2017). Biochar as an adsorbent for inorganic nitrogen and phosphorus removal from water: A review. *Environ. Sci. Pollut. Res.,* 24, pp. 26297–26309.

Yu, H., Shang, L., Bian, T., Shi, R., Waterhouse, G. I. N., Zhao, Y., Zhou, C., Wu, L.-Z., Tung, C.-H. and Zhang, T. (2016). Nitrogen-Doped porous carbon nanosheets templated from g-C_3N_4 as metal-free electrocatalysts for efficient oxygen reduction reaction, *Adv. Mater.*, 28, pp. 5080–5086.

Yu, L., Falco, C., Weber, J., White, R. J., Howe, J. Y. and Titirici, M.-M. (2012). Carbohydrate-derived hydrothermal carbons: A thorough characterization study. *Langmuir*, 28, pp. 12373–12383.

Zhang, C., Cheng, Z., Fu, Z., Liu, Y., Yi, X., Zheng, A., Kirk, S. R. and Yin, D. (2017). Effective transformation of cellulose to 5-hydroxymethylfurfural catalyzed by fluorine anion-containing ionic liquid modified biochar sulfonic acids in water. *Cellulose*, 24, pp. 95–106.

Zhang, F., Wang, X., Yin, D., Peng, B., Tan, C., Liu, Y., Tan, X. and Wu, S. (2015). Efficiency and mechanisms of Cd removal from aqueous solution by biochar derived from water hyacinth (*Eichornia crassipes*). *J. Environ. Manage.*, 153, pp. 68–73.

Zhao, L., Bacsik, Z., Hedin, N., Wei, W., Sun, Y., Antonietti, M. and Titirici, M.-M. (2010). Carbon dioxide capture on amine-rich carbonaceous materials derived from glucose. *ChemSusChem*, 3, pp. 840–845.

Zhao, R. and Zhang, R.-Q. (2017). Beyond the electrostatic model: the significant roles of orbital interaction and the dispersion effect in aqueous–π systems. *Phys. Chem. Chem. Phys.*, 19, pp. 1298–1302.

Chapter 7

Biochar-based Carbon Materials for Energy-Storage Applications

Niklas Hedin[*] and Peng Zhang

*Department of Materials and Environmental Chemistry,
Stockholm University, SE 106 91 Stockholm, Sweden*

*niklas.hedin@mmk.su.se

Abstract

Energy-storage is important if the world's energy system is to become less dependent on fossil fuels. Such storage involves e.g. the storage of heat or charged ions, and carbon-rich materials derived from biochars are relevant. In this chapter, we review the use of mainly activated carbons (ACs) derived from hydrochars or pyrolytic biochars for heat storage, and as electrode materials for supercapacitors. Suggestions for further studies are being made.

7.1 Introduction

Attenuated human consumption of energy is necessary for a modern and high quality life, but it is connected with significant risks in relation to global warming. About 82% of the energy supply of the world of 2015 was based on fossil fuels, and the final energy consumption amounted to $3.9 \cdot 10^{20}$ J, which corresponded to an average power use of 1.7 kW/ person. The emissions of CO_2 from fossil fuels were 32300 Mt [IEA, 2017]. Those emissions are connected with global warming and sea level rise, and in turn are related to significant risks to the human population,

165

with undernutrition, vector-borne diseases, decline in occupational and mental health, violence, reduced air quality, food and water borne infections and heat [IPCC, 2014]. These risks have led to the policy makers, companies and researchers focusing significant efforts towards reducing the fossil-fuel usage and adapting to the effects of global warming.

To increase the fraction of nuclear energy from today's 5% of the energy supply is costly and potentially risky, and efforts are directed towards increasing the supply of bioenergy, hydropower and solar power. Wind and solar power are getting increasing attention, although they are only a minor part of the overall energy supply, but their contributions are rapidly increasing. With a significantly reduced usage of fossil fuels, the temporal variations in the energy supply, over multiple time scales, will increase. It is with respect to these variations, and the electrification of the transport sector, that the enhanced need for energy storage should be understood.

The overall class of carbon-rich materials of this book, including biochars and related activated carbons (ACs), are relevant to energy storage. As a kind of charcoal or peat, biochars have many advantages over other carbon-rich materials, including low cost, high stability, nontoxicity and high sustainability [Manyà, 2012]. Biochars have been used for centuries and can be made from waste biomass (such as agricultural crops, seeds, grass, wild plants, trees, etc.) via pyrolysis, or hydrothermal carbonization processes, or gasification. The International Biochar Initiative has defined a biochar as "a solid material obtained from the thermochemical conversion of biomass in an oxygen-limited environment" [IBI, 2015]. The three methods commonly used for the synthesis or production of biochars have different yields [Qian *et al.*, 2015], and for gasification it is low, so we regard pyrolysis and hydrothermal carbonization as the main methods to obtain biochars [Tripathi *et al.*, 2016]. Biochars can contain oxygenated groups, such as carboxyl, hydroxyl and phenolic functional groups [Cao and Harris, 2010; Zhu *et al.*, 2015], and we call them "pyrolytic biochar" or "hydrochar" when produced by pyrolysis or hydrothermal methods, respectively.

Most of the applications of biochars in the field of energy storage have involved the use of ACs. ACs are extremely porous, have a large surface area, and are inexpensive. They can be prepared from both pyrolytic biochars or hydrochars [Cha *et al.*, 2016]. ACs typically have specific surface areas of 500–3000 m^2/g, pore volumes of 0.5–1.5 cm^3/g, and can be prepared by different means. ACs are commonly described with the general method of preparation as being "physically activated" or "chemically activated", but both approaches certainly involve chemical reactions. The surfaces of ACs contain oxygenated compounds rendering them with acid or base characters. They are black, conduct heat and electricity relatively well, which are relevant for storage of heat and charged ions.

Energy can be stored in many ways that include gravitational or mechanical storage, storage of heat and storage of charges (electrons or ions) [Amrouche *et al.*, 2016]. Gravitational storage is appropriate for long-term storage, and mechanical storage is established for shorter-term storage, and both approaches are less relevant to this book chapter. Heat can be stored by sensible and latent heat and chemical reactions [Zalba *et al.*, 2003]. Biochar-based carbon materials offer advantages to heat storage but have not yet been extensively researched with respect to this. When it comes to storage of charged ions and electrons, much more research has been performed, mainly in association with supercapacitors but also in relation to batteries. Supercapacitors store high-power electrochemical energy and have very high capacitances, long cycle life, and a relatively wide range of operational temperatures [Yu *et al.*, 2013]. Today they are used in many power-management applications that require rapid charge–discharge cycles for short-term power needs, such as in backup power and medical devices. Their applications are limited by the relatively low energy densities, and research is being directed into increasing the energy densities [Candelaria *et al.*, 2012]. Supercapacitors are being researched for applications in vehicles, trains, fluctuation stabilizers in the electrical net, and charging stations for electrical vehicles [Yu *et al.*, 2013].

As mentioned, heat can be stored as sensible and latent heat, and chemical reactions [Zalba *et al.*, 2003]. Liu *et al.* [2018] recently studied the effect of adding biochar to soil and observed that it could either have a negative effect on the thermal properties of soil, as it increased the porosity, or have a positive effect as it increased the solid water content. In addition, ACs are suitable sorbents in working pairs with ammonia or methanol for adsorptive heat storage in relation to adsorptive refrigeration.

7.2 Phase Change Materials

Heat storage can be conducted with phase change materials (PCMs) able to store and release latent heat on melting and solidification. A long range of inorganic, organic PCMs have been studied, and some salt hydrates, salt solutions and paraffin compositions have been commercialized [Zalba *et al.*, 2003]. Microencapsulation of PCMs can prevent leakage of the liquid phase and reduce the corrosion problems associated with salt solutions and hydrates [Jamekhorshid *et al.*, 2014]. For example, a succulent-based aerogel based on a pyrolized spongy tissue was prepared and effectively used as a substrate for a paraffin-based PCM [Wei *et al.*, 2018]. Particles of a glucose-derived hydrochar were used to prepare a composite with a PCM (polyethylene glycol), which was studied thermally [Sun *et al.*, 2017]. Carbonized and KOH-activated chitosan was also studied as a support for a PCM (hexadecanol) [Fang *et al.*, 2017]. ACs prepared from a hemp-derived hydrochar were added to enhance the thermal conductivity of a eutectic mixture of oleic and capric acid, and the resulting mixtures were studied thermally at low temperatures [Hussain *et al.*, 2017]. An NaOH-activated pyrolitic biochar was studied as an AC together with graphene nanoplates with respect to the thermal conductivity enhancement of a PCM (palmitic acid) [Mehrali *et al.*, 2016]. For PCMs within small pores of carbons, the melting point is expected to be depressed by a Kelvin effect [Alba-Simionesco *et al.*, 2006], which was also shown for a glucose-derived carbon aerogel filled with a PCM (octadecanol) [Huang *et al.*, 2014].

7.3 Adsorptive Heat Storage

Adsorptive systems for refrigerators, air-conditioning and heat pumps have some advantages over regular and photovoltaic vapor-compressive refrigeration. The main advantage is that adsorptive refrigeration does not make use of Freons^{TM} or similar molecules. Note that hotter heat sources can also be used in adsorptive refrigeration than in traditional absorptive systems [Ullah *et al.*, 2013]. However, this kind of refrigeration is still not competitive today and further studies need to be performed to increase their performance and reduce their overall cost [Yeo *et al.*, 2012]. Further research into carbon-based adsorbents and the potential to replace the fluids (commonly used fluids are NH_3 and methanol) to less toxic ones is warranted.

Oil-palm-shell-derived ACs have been studied for this application as they have good properties and low cost, and Abdullah *et al.* reviewed their use in relation to air-conditioning system for vehicles [Abdullah *et al.*, 2011]. Pal *et al.* [2017] prepared pyrolytic biochars from waste-palm trunk, and mangrove, and activated them into high-surface area ACs with KOH. These ACs were studied with respect to the adsorption of ethanol, and the heat of adsorption. It was concluded that they would likely perform better than the commercial sorbent Maxsorb III AC.

7.4 Biochars and ACs as Electrode Materials for Supercapacitor

Biochar-derived carbon materials are being intensely researched for the storage of charges of ions and electrons. The research involves studies of electrode materials for supercapacitors, and lithium-ion and other types of batteries. This chapter limits the discussion to that of supercapacitors, and should not be seen as a full review.

7.4.1 *Electrostatic double layer capacitors*

In electrostatic double layer capacitors (EDLCs), ACs and other high-surface area carbons are used as electrode materials, as they provide a very large specific surface area, and a very short equivalent distance among the plates in the capacitor. The distance will be of the order of a few Ångströms. A high surface area, a pore size distribution (PSD) matched with respect to the size of the ions and the solvent molecules, and a high conductivity are appreciated [Yu *et al.*, 2013]. Commercial ACs for EDLCs are today mainly produced from pyrolytic biochars from coconut shells as they have good properties and a low cost. It is important to observe that some EDLCs operate with aqueous electrolytes (with acids, bases or salts to enhance the conductivity of the solution), others with organic electrolytes (with quaternary ammonium salts, typically with BF_4^- or PF_6^- anions, in solvents such as acetonitrile), and ionic liquids are also researched as electrolytes. For aqueous electrolytes, significantly smaller potentials need to be used as compared to the non-aqueous ones, which in turn lead to a lower power capacity. The specific capacitance is typically somewhat higher for aqueous electrolytes as compared to the non-aqueous ones, as the latter ones have larger ions [Yu *et al.*, 2013].

Biochars are suitable starting materials for ACs designed for EDLCs. Selecting the proper biomass and pyrolysis or hydrothermal carbonization conditions allow for a detailed control of the specific surface area and the PSD [Chun *et al.*, 2011; Wei *et al.*, 2011]. The biochars contain a large fraction of oxygenated functional groups, e.g. –OH, –CHO and –COOH, which can be adjusted by the process conditions such as the temperature used for the activation [Jiang *et al.*, 2013; Jin *et al.*, 2013; Gupta *et al.*, 2015]. Note that the heteroatoms in O- [Frackowiak, 2007] and N-containing groups [Deng *et al.*, 2016] can enhance the pseudocapacitance of the EDLCs. However, as the O-containing groups seem to be related to observed long-term decay of the capacitance [Azais *et al.*, 2007], it is not clear if this kind of pseudocapacitance can be effectively harvested in practical applications of supercapacitors.

The composition of the hydrochars is controlled by the hydrothermal chemistry (via the temperature, time of reaction, concentration of reactants, the precursor type, etc.) [Titirici *et al.*, 2008]. The composition influences the properties of the ACs, but the largest dependencies are related to the details of the activation processes. The capacitances of the ACs benefit from high porosities and surface areas [Kondrat *et al.*, 2012], and a proper activation of the hydrochar is needed to assure an increased surface area and porosity. Both physical activation and chemical activation have been used in numerous studies to increase the surface area and porosity of the ACs prepared from hydrochars (and pyrolytic biochars) [Falco *et al.*, 2013; Deng *et al.*, 2014; Manyà *et al.*, 2018; Hao *et al.*, 2013; Sevilla *et al.*, 2017].

Pyrolytic biochars typically contain less O- and N-atoms than hydrochars [Eibisch *et al.*, 2015]. As for hydrochars [Titirici and Antonietti, 2010], the pyrolysis and activation procedures, and the choice of initial biomass [Tripathi *et al.*, 2016] have been shown to be important to the properties of the ACs, which in turn affects their properties as electrode materials in supercapacitors.

It is also important to note that only a few factories exist for the synthesis of ACs with KOH, and research is being conducted towards finding alternative activation agents [Sevilla *et al.*, 2017]. Usually, the KOH-chemical activation method is widely used for activation of biochars in the open literature, but also $KHCO_3$-activations are being studied [Basta *et al.*, 2009; Sevilla and Fuertes, 2016].

Lee *et al.* [2016] prepared ACs from hydrochars of sugars and chemical activation with KOH. The hydrochars and the activation processes were optimized to tailor the properties (e.g. textural properties, chemical composition, N-doping, electrical conductivity) of the ACs. The Brunauer–Emmett–Teller (BET) surface areas could be tuned in the range of 800–3000 m^2/g with associated variation in the extent of microporosity and

PSD. At the same time, it was confirmed that the PSDs could be controlled by the activation temperature. At temperatures ≤700°C, the ACs had narrow PSDs with predominantly ultramicropores; and at >700°C, larger pores were formed. As electrodes for supercapacitors, a high specific capacitance (C_m) of ~260 F/g was achieved at a moderately high S_{BET} of ~1300 m²/g, which was equivalent to a C_m/S_{BET} of 20 μF/cm², for an optimal AC prepared from a hydrochar of glucose. Cyclic voltammetry (CV) curves and plots for the capacitance as a function of the scan rate are presented in Figure 7.1 for ACs prepared from hydrochars from glucosamine (GA) with

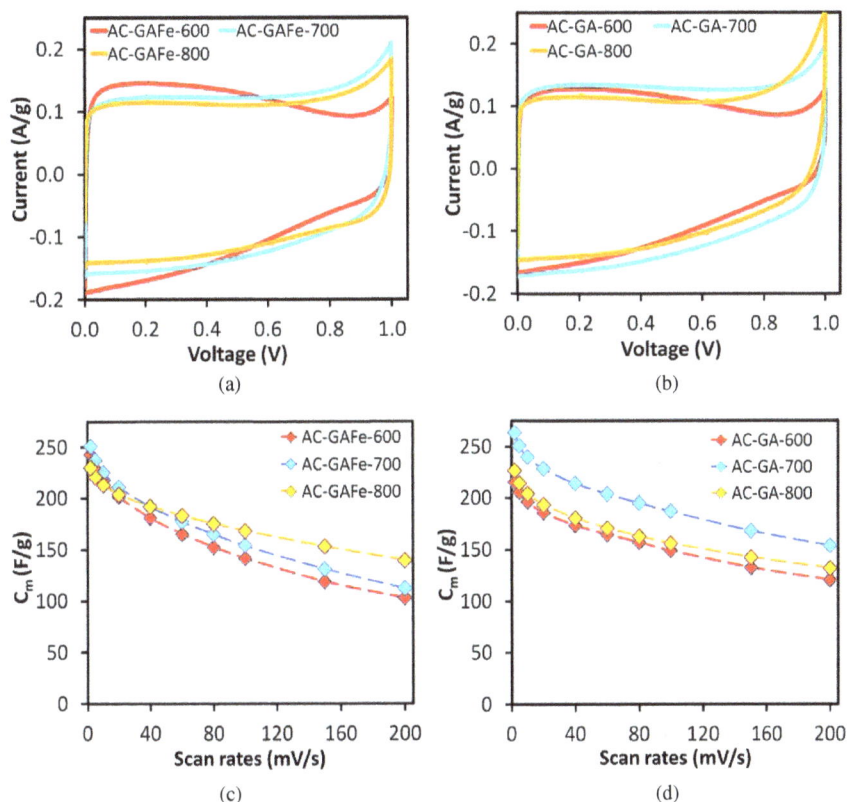

Figure 7.1: (a) and (b) *CV* curves of different ACs prepared from hydochars of glucosamine at 2 mV s⁻¹, and ((c) and (d)) specific capacitances (C_m) calculated from *CV* scanned from 2 to 200 mV s⁻¹. AC-X-Y denotes (X: GA or GAFe) prepared at temperatures of Y: 600–800°C by activation with KOH.

Source: Reprinted with permission from Lee *et al.* [2016].

and without iron oxide. Similar studies of hydrochars from various sugars and biomass, and associated ACs have been studied by numerous authors. In those studies, similar capacitances have been observed for ACs prepared from hydrochars and pyrolytic biochars [Wei *et al.*, 2011; Jiang *et al.*, 2013; Jin *et al.*, 2013; Gupta *et al.*, 2015; Basta *et al.*, 2009; Sevilla and Fuertes, 2016; Goldfarb *et al.*, 2017; Sevilla *et al.*, 2014; Fuertes and Sevilla, 2015; Zheng *et al.*, 2014; Zhao *et al.*, 2010].

From a textbook perspective, the specific surface area would be important for the EDLCs, but also the micropore volume is expected to tune the effective distance in the supercapacitors. Recent findings have although shown that there is no simple scaling in between the BET surface area and the capacitance, and optimizing the textural properties of the AC is beneficial [Borchardt *et al.*, 2014]. An appropriately snug fit of the ions from the electrolyte in the pore structure seems to be advantageous [Chmiola *et al.*, 2006], and different optima are expected for the anode and cathode electrode materials. The study of Lee *et al.* [2016] showed quite well-defined scaling relationships between the micropore volume derived from the t-plot method and the capacitances observed, see Figure 7.2. Centeno and Stoeckli [2006] showed that the micropore specific surface area was an appropriate predictor of the specific capacitance in their study of the enhanced capacitance of O-rich ACs.

It has been investigated and suggested that microporous–mesoporous structures are necessary or preferred for an excellent performance of electrode materials of supercapacitors [Farma *et al.*, 2013]. Mesopores could enhance the mass transport of ions and in turn the power capacity of the supercapacitors. Dehkhoda *et al.* [2016] studied the effect of introducing mesoporosity in the ACs derived from pyrolytic biochar and KOH activation and concluded that the electrode resistance was reduced and that the capacitive behavior was improved. Sun *et al.* [2016] studied ACs derived from pyrolytic biochar from hemp and concluded that the mesopores improved on the capacitance observed.

N-doped carbons are actively being researched as electrode material in supercapacitors because of the electronic and hydrophilic properties of the framework [Deng *et al.*, 2016]. They are typically denser than those without N-atoms [Frackowiak, 2007]. It has been suggested that an N-enriched AC would be beneficial to the supercapacitance by tuning the conductivity of the carbon, contributions by pseudo-capacitance or the density. Numerous types of N-enriched ACs have been prepared from hydrochars [Wei *et al.*, 2011; Zhang and Wu, 2014; Lee *et al.*, 2018; Ma *et al.*, 2017; Si *et al.*, 2013], and pyrolytic biochars [Demir *et al.*, 2018; Raj *et al.*, 2018;

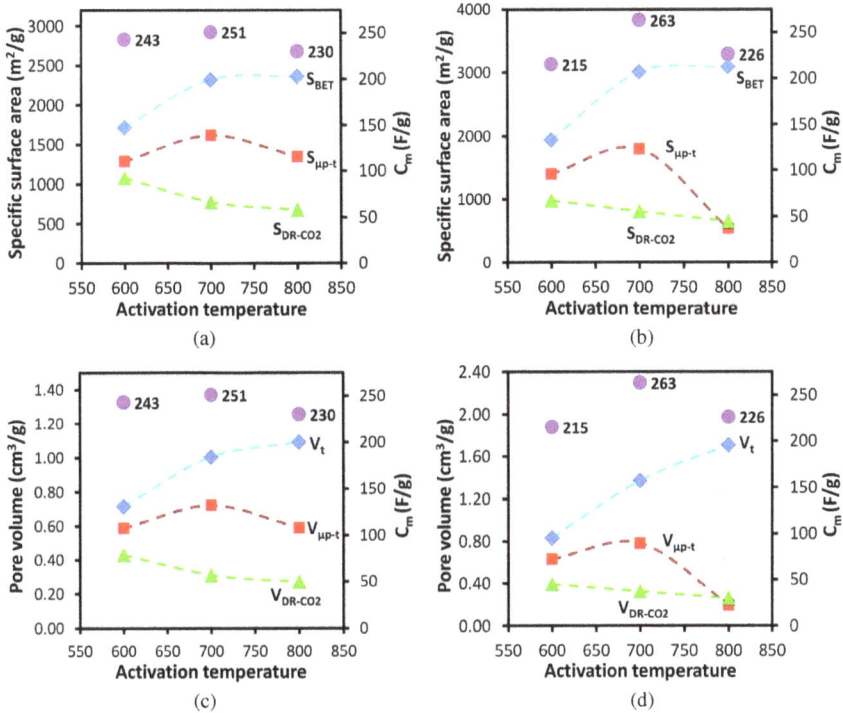

Figure 7.2:   Relationships between the specific surface areas and pore volumes (dashed lines) calculated by various methods and the capacitances Cm (purple dots) for AC-GAFe ((a) and (c)) and AC-GA ((b) and (d)). Activated carbon (AC-X) from hydrochars from glucosamine with and without iron (X: GA or GAFe).

Source: Reprinted with permission from Lee *et al.* [2016].

Kim *et al.*, 2017; Gao *et al.*, 2017; Niu *et al.*, 2017]. For example, Demir *et al.* [2018] studied N- and O-doped ACs from activated pyrolytic pea protein and observed high capacitances (413 F/g at 1.0 A/g in 1 M of KOH) amplified by pseudocapacitance and ascribed the latter to contributions from N- and O-containing moieties. At the current stage of development, it is difficult to access if the N-rich carbons and the pseudocapacitance can be used in devices or if they would impose long-term stability issues [Azais *et al.*, 2007].

The electrical conductivities of ACs are significantly smaller than those of metals and improving the conductivities could reduce the Ohmic losses occurring in supercapacitors [Yu *et al.*, 2013]. Lee *et al.* [2016] studied the influence of the hydrothermal conditions of hydrochars with respect to the conductivities of the resulting KOH-activated carbons

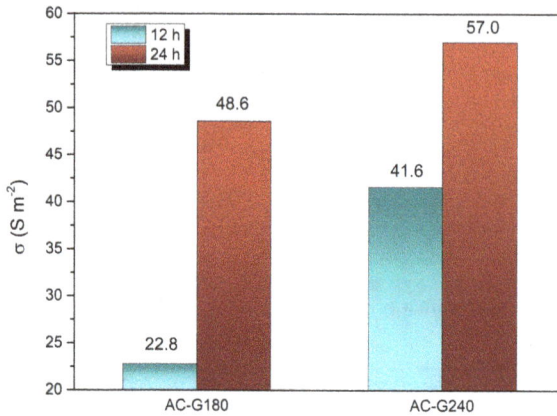

Figure 7.3:   Electrical conductivities for ACs prepared from hydrochars from glucose synthesized at different conditions (temperature: 180°C (G180), 240°C (G240); duration: 12 and 24 h). The activation was performed with KOH.

Source: Reprinted with permission from Lee *et al.* [2016].

(KOH-ACs). It was observed that the hydrothermal temperature (180°C vs. 240°C) had only a minor influence on the textural properties of the KOH-ACs. However, the hydrothermal conditions did affect the electrical conductivity of the KOH-ACs. A comparably high conductivity was observed for the ACs from hydrochars prepared at a high temperature and a short time for hydrothermal carbonization, as can be deduced from the data in Figure 7.3. Overall, these tendencies seem to argue that hydrochars should be prepared at a relatively high temperature for this application, or that pyrolytic biochars should be used. Further studies into why a short time for the hydrothermal carbonization was preferred with respect to the conductivities of the related ACs would be needed. One could also contemplate performing studies of introducing a pyrolysis step before the activation of the hydrochars to achieve ACs with the high conductivity being important to supercapacitors.

7.4.2 *Hybrid supercapacitors*

In addition to EDLCs, supercapacitors can be based on chemical reactions, and also hybrids of the chemical supercapacitors and EDLCs are being researched [IPCC, 2014]. Several of these hybrid supercapacitors have been derived, partly, from activated biochars. Genovese *et al.* studied ACs prepared by KOH-activation of pyrolytic biochars from pine cones and adsorbed redox-active polyoxometalate clusters, and these hybrid materials presented an 2.5-fold enhanced areal capacitance as compared with the corresponding ACs [Genovese and Lian, 2017]. Wang *et al.* [2017] studied ACs derived from pyrolytic biochars and observed a significantly enhanced capacitance when the ACs had been hybridized with Ni/NiO/NiOOH.

7.5 Conclusions

The suitability of sustainable carbon materials derived from hydrochars or pyrolytic biochars for energy storage was reviewed. The focus was held on applications of such carbon materials in the storage of heat in charged ions and electrons. For heat storage, PCMs can be encapsulated and contained within the porous carbons and also enhance the thermal conductivity. ACs, derived from biochars, are also proving to be relevant sorbents in the adsorptive refrigeration technology. It is still uncertain if adsorptive refrigeration can compete with commercial means to refrigeration. Still, the ACs derived from biochars seem to have at least as good properties as commercialized variants of ACs, and it would be interesting to know if less toxic fluids than NH_3 or methanol could be effectively used.

Electrode materials for EDLCs derived from biochars offer good sustainability and a low cost. In our view, it is likely that ACs, in general, will be less expensive than more exotic materials, such as graphene, for use in electrode materials in EDLC, and also be more sustainable. The long-term stability of supercapacitors from ACs demands more studies, especially if pseudocapacitances or hybrid materials should be used. It seems as though the hydrochars should either be prepared at a relatively high temperature or be pyrolyzed in an intermediate step before the activation to assure sufficiently high conductivity. In addition, even if the ACs in the open literature are usually being synthesized by KOH activation, concerns should be noted when it comes to the scalability of this activation procedure. Developments of alternative methods for the activation might be warranted.

References

Abdullah, M. O., Tana, I. A. W. and Lim, L. S. (2011). Automobile adsorption air-conditioning system using oil palm biomass-based activated carbon: A review. *Renew. Sust. Energy Rev.*, 15, pp. 2061–2072.

Alba-Simionesco, C., Coasne, B., Dosseh, G., Dudziak, G., Gubbins, K. E., Radhakrishnan, R. and Sliwinska-Bartkowiak, M. (2006). Effects of confinement on freezing and melting. *J. Phys. Cond. Mat.*, 18, pp. R15–R68.

Amrouche, S. O., Rekioua, D., Rekioua, T. and Bacha, S. (2016). Overview of energy storage in renewable energy. *Int. J. Hyd. Energy*, 41, pp. 20914–20927.

Azais, P., Duclaux, L., Florian, P., Massiot, D., Lillo-Rodenas, M.-A., Linares-Solano, A., Peres, J.-P., Jehoulet, C. and Beguin, F. (2007). Causes of supercapacitors ageing in organic electrolyte. *J. Power Sources*, 171, pp. 1046–1053.

Basta, A. H., Fierro, V., El-Saied, H. and Celzard, A. (2009). 2-Steps KOH activation of rice straw: An efficient method for preparing high-performance activated carbons. *Bioresour Technol.*, 100, pp. 3941–3947.

Borchardt, L., Oschatz M. and Kaskel, S. (2014). Tailoring porosity in carbon materials for supercapacitor applications. *Mater. Horiz.*, 1, pp. 157–168.

Candelaria, S. L., Shao, Y. Y., Zhou, W., Li, X. L., Xiao, J. Zhang, J. G., Wang, Y., Liu, J., Li, J. H. and Cao, G. Z. (2012). Nanostructured carbon for energy storage and conversion. *Nano Energy*, 1, pp. 195–220.

Centeno, T. A. and Stoeckli, F. (2006). The role of textural characteristics and oxygen-containing surface groups in the supercapacitor performances of activated carbons. *Electrochim. Acta*, 52, pp. 560–566.

Cha, J. S., Park, S. H., Jung, S.-C., Ryu, C., Jeon, J.-K., Shin, M.-C. and Park, Y.-K. (2016). Production and utilization of biochar: A review. *J. Ind. Eng. Chem.*, 40, pp. 1–15.

Cao, X. and Harris, W. (2010). Properties of dairy-manure-derived biochar pertinent to its potential use in remediation. *Bioresource Technol.*, 101, pp. 5222–5228.

Chmiola, J., Yushin, G., Gogotsi, Y., Portet, C., Simon P. and Taberna, P. L. (2006). Anomalous increase in carbon capacitance at pore sizes less than 1 nanometer. *Science*, 313, pp. 1760–1763.

Chun, S.-E., Picard, Y. N. and Whitacre, J. F. (2011). Relating precursor pyrolysis conditions and aqueous electrolyte capacitive energy storage properties for activated carbons derived from anhydrous glucose-d. *J. Electrochem. Soc.*, 158, A83–A92.

Dehkhoda, A. M., Gyenge, E. and Ellis, N. (2016). A novel method to tailor the porous structure of KOH-activated biochar and its application in capacitive deionization and energy storage. *Biomass Bioenergy*, 87, pp. 107–121.

Demir, M., Ashourirad, B., Mugumya, J. H., Saraswat, S. K., El-Kaderi, H. M. and Gupta, R. B. (2018). Nitrogen and oxygen dual-doped porous carbons prepared from pea protein as electrode materials for high performance super-capacitors, in press. *Int. J. Hydrogen Energy*, https://doi.org/10.1016/j. ijhydene.2018.03.220.

Deng, S. B., Wei, H. R., Chen, T., Wang, B., Huang, J. and Yu, G. (2014). Superior CO_2 adsorption on pine nut shell-derived activated carbons and the effective micropores at different temperatures. *Chem. Eng. J.*, 253, pp. 46–54.

Deng, Y., Xie, Y., Zou, K. and Ji, X. (2016). Review on recent advances in nitrogen-doped carbons: Preparations and applications in supercapacitors. *J. Mater. Chem. A.*, 4, pp. 1144−1173.

Eibisch, N., Schroll, R., Fuss, R., Mikutta, R., Helfrich, M. and Flessa, H. (2015). Pyrochars and hydrochars differently alter the sorption of the herbicide iso-proturon in an agricultural soil. *Chemosphere*, 119, pp. 155–162.

Falco, C., Marco-Lozar, J. P., Salinas-Torres, D., Morallon, E., Cazorla-Amoros, D., Titirici, M. M. and Lozano-Castello, D. (2013). Tailoring the porosity of chemically activated hydrothermal carbons: Influence of the precursor and hydrothermal carbonization temperature, *Carbon*, 62, pp. 346–355.

Fang, X., Hao, P., Song, B., Tuan, C.-C., Wong, C.-P. and Yu, Z.-T. (2017). Form-stable phase change material embedded with chitosan-derived carbon aero-gel. *Mater. Lett.*, 195, pp. 79–81.

Farma, R., Deraman, M., Awitdrus A., Talib I. A., Taer, E., Basri N. H., Manjunatha, J. G., Ishak, M. M., Dollah, B. N. M. and Hashmi, S. A. (2013). Preparation of highly porous binderless activated carbon electrodes from fibres of oil palm empty fruit bunches for application in supercapacitors. *Bioresource Technol.*, 132, pp. 254–261.

Frackowiak, E. (2007). Carbon materials for supercapacitor application. *Phys. Chem. Chem. Phys.*, 9, pp. 1774–1785.

Fuertes A. B. and Sevilla, M. (2015). Superior capacitive performance of hydro-char-based porous carbons in aqueous electrolytes. *ChemSusChem*, 8, pp. 1049–1057.

Genovese, M. and Lian, K. (2017). Polyoxometalate modified pine cone biochar carbon for supercapacitor electrodes, *J. Mater. Chem. A*, 5, pp. 3939–3947.

Goldfarb, J. L., Dou, G. L., Salari, M. and Grinstaff, M. W. (2017). Biomass-based fuels and activated carbon electrode materials: An integrated approach to green energy systems. *ACS Sustainable Chem. Eng.*, 5, pp. 3046−3054.

Guo, D., Zheng, C., Deng, W., Chen, X., Wei, H., Liu, M. and Huang, S. (2017). Nitrogen-doped porous carbon plates derived from fallen camellia flower for electrochemical energy storage. *J. Solid State Electrochem.*, 21, pp. 1165−1174.

Gupta, R. K., Dubey, M., Kharel, P., Gu, Z. and Fan, Q. H. (2015). Biochar activated by oxygen plasma for supercapacitors. *J. Power Sources*, 274, pp. 1300–1305.

Hao, W. M., Bjorkman, E., Lilliestråle, M. and Hedin, N. (2013). Activated carbons prepared from hydrothermally carbonized waste biomass used as adsorbents for CO_2. *Applied Energy*, 112, pp. 526–532.

Huang, X. Y., Xia, W. and Zou, R. Q. (2014). Nanoconfinement of phase change materials within carbon aerogels: Phase transition behaviours and photo-to-thermal energy storage. *J. Mater. Chem. A.*, 2, pp. 19963–19968.

Hussain, S. I., Dinesh, R., Roseline, A. A., Dhivya, S. and Kalaiselvam, S. (2017). Enhanced thermal performance and study the influence of sub cooling on activated carbon dispersed eutectic PCM for cold storage applications. *Energy Buildings*, 143, pp. 17–24.

IBI (The International Biochar Initiative) (2015). Standardized product definition and product testing guidelines for biochar that is used in soil, IBI-STD-2.1, (downloaded September 22, 2018 from https://biochar-international.org/characterizationstandard/).

IEA (International Energy Agency) (2017). Key World Energy Statistics.

IPCC (United Nations Intergovernmental Panel on Climate Change) (2014). The fifth assessment report (AR5).

Jamekhorshid, A., Sadrameli, S. M. and Farid, M. (2014). A review of microencapsulation methods of phase change materials (PCMs) as a thermal energy storage (TES) medium. *Renew. Sust. Energy Rev.*, 31, pp. 531–542.

Jiang, J. H., Zhang, L., Wang, X. Y., Holm, N., Rajagopalan, K., Chen, F. L. and Ma, S. G. (2013). Highly ordered macroporous woody biochar with ultra-high carbon content as supercapacitor electrodes. *Electrochim. Acta.*, 113, pp. 481–489.

Jin, H., Wang, X. M., Gu, Z. R. and Polin, J. (2013). Carbon materials from high ash biochar for supercapacitor and improvement of capacitance with HNO_3 surface oxidation. *J. Power Sources*, 236, pp. 285–292.

Kim, C. K., Choi, I. T., Kang, S. H. and Kim, H. K. (2017). Anchovy-derived nitrogen and sulfur co-doped porous carbon materials for high-performance supercapacitors and dye-sensitized solar cells. *RSC Adv.*, 7, pp. 35565–35574.

Kondrat, S., Pérez, C. R., Presser, V., Gogotsi, Y. and Kornyshev, A. A. (2012). Effect of pore size and its dispersity on the energy storage in nanoporous supercapacitors. *Energy Environ. Sci.*, 5, 6474–6479.

Lee, K. K., Björkman, E., Morin, D., Lilliestråle, M., Björefors, F., Andersson, A. M. and Hedin, N. (2016). Effects of hydrothermal carbonization conditions on the textural and electrical properties of activated carbons. *Carbon*, 107, pp. 619–621.

Lee, K. K., Hao, W. M., Gustafsson, M., Tai, C.-W., Morin, D., Björkman, E., Lilliestråle, M., Björefors, F., Andersson, A. M. and Hedin, N. (2016).

Tailored activated carbons for supercapacitors derived from hydro-thermally carbonized sugars by chemical activation. *RSC Adv.*, 6, pp. 110629–110641.

Lee, K. K., Church, T. L. and Hedin, N. (2018). RNA as a Precursor to N-Doped Activated Carbon. *ACS Appl. Energy Mater.*, 1, pp. 3815−3825.

Liu, Z. P., Xu, J. N., Li, X. L. and Wang, J. F. (2018). Mechanisms of biochar effects on thermal properties of red soil in south China. *Geoderma*, 323, pp. 41–51.

Ma, H., Li, C., Zhang, M., Hong, J.-D. and Shi, G. (2017). Graphene oxide induced hydrothermal carbonization of egg proteins for high-performance supercapacitors. *J. Mater. Chem. A*, 5, pp. 17040−17047.

Manyá, J. J. (2012). Pyrolysis for biochar purposes: A review to establish current knowledge gaps and research needs. *Environ. Sci. Technol.*, 46, pp. 7939−7954.

Manyà, J. J., Gonzalez, B., Azuara, M. and Arner, G. (2018). Ultra-microporous adsorbents prepared from vine shoots-derived biochar with high CO_2 uptake and CO_2/N_2 selectivity. *Chem. Eng. J.*, 345, pp. 631–639.

Mehrali, M., Latibari, S. T., Rosen, M. A., Akhiani, A. R., Naghavi, M. S., Sadeghinezhad, E., Metselaar, H. S. C., Nejad, M. M. and Mehrali, M. (2016). From rice husk to high performance shape stabilized phase change materials for thermal energy storage. *RSC Adv.*, 6, pp. 45595–45604.

Niu, Q., Gao, K., Tang, Q., Wang, L., Han, L., Fang, H., Zhang, Y., Wang, S. and Wang, L. (2017). Large-size graphene-like porous carbon nanosheets with controllable N-Doped surface derived from sugarcane bagasse pith/chitosan for High Performance Supercapacitors, *Carbon*, 123, pp. 290−298.

Pal, A., Thu, K., Mitra, S., El-Sharkawy, I. I., Saha, B. B., Kil, H.-S., Yoon, S.-H. and Miyawaki, J. (2017). Study on biomass derived activated carbons for adsorptive heat pump application. *Int. J. Heat Mass Trans.*, 110, pp. 7–19.

Qian, K., Kumar, A., Zhang, H., Bellmer, D. and Huhnke, R. (2015). Recent advances in utilization of biochar. *Renew. Sust. Energy Rev.*, 42, pp. 1055–1064.

Raj, C. J., Rajesh, M., Manikandan, R., Yu, K. H., Anusha, J. R., Ahn, J. H., Kim, D.-W., Park, S. Y. and Kim, B. C. (2018). High electrochemical capacitor performance of oxygen and nitrogen enriched activated carbon derived from the pyrolysis and activation of squid gladius chitin. *J. Power Sources*, 386, pp. 66−76.

Sevilla, M., Yu, L., Ania, C. O. and Titirici, M.-M. (2014). Supercapacitive behavior of two glucose-derived microporous carbons: Direct pyrolysis versus hydrothermal carbonization. *ChemElectroChem*, 1, pp. 2138–2145.

Sevilla, M. and Fuertes, A. B. (2016). A green approach to high-performance supercapacitor electrodes: The chemical activation of hydrochar with potassium bicarbonate. *ChemSusChem*, 9, pp. 1880–1888.

Sevilla, M.; Ferrero, G. A and Fuertes, A. B. (2017). Beyond KOH activation for the synthesis of superactivated carbons from hydrochar. *Carbon*, 114, pp. 50–58.

Si, W., Zhou, J., Zhang, S., Li, S., Xing, W. and Zhuo, S. (2013). Tunable N-Doped or dual N, S-Doped activated hydrothermal carbons derived from human hair and glucose for supercapacitor applications. *Electrochim. Acta*, 107, pp. 397–405.

Sun, Q. R., Yuan, Y. P., Zhang, H. Q., Cao, X. L. and Sun, L. L. (2017). Thermal properties of polyethylene glycol/carbon microsphere composite as a novel phase change material. *J. Therm. Anal. Calorim.*, 130, pp. 1741–1749.

Sun, W., Lipka, S. M., Swartz, C., Williams, D. and Yang, F. Q. (2016). Hemp-derived activated carbons for supercapacitors. *Carbon*, 103, pp. 181–192.

Titirici, M. M., Antonietti, M. and Baccile, N. (2008). Hydrothermal carbon from biomass: A comparison of the local structure from poly- to monosaccharides and pentoses/hexoses. *Green Chem.*, 10, pp. 1204–1212.

Titirici, M. M. and Antonietti, M. (2010). Chemistry and materials options of sustainable carbon materials made by hydrothermal carbonization. *Chem. Soc. Rev.*, 39, pp. 103–116.

Tripathi, M., Sahu, J. N. and Ganesan, P. (2016). Effect of process parameters on production of biochar from biomass waste through pyrolysis: A review. *Renew. Sust. Energy Rev.*, 55, pp. 467–481.

Ullah, K. R., Saidur, R., Ping, H. W., Akikur, R. K. and Shuvo, N. H. (2013). A review of solar thermal refrigeration and cooling methods. *Renew. Sust. Energy Rev.*, 24, pp. 499–513.

Wang, Y. F., Zhang, Y., Pei, L., Ying, D. W., Xu, X. Y., Zhao, L., Jia, J. P. and Cao, X. D. (2017). Converting Ni-loaded biochars into supercapacitors: Implication on the reuse of exhausted carbonaceous sorbents. *Sci. Rep.*, 7, p. 41523.

Wei, L., Sevilla, M., Fuertes, A. B., Mokaya, R. and Yushin, G. (2011). Hydrothermal carbonization of abundant renewable natural organic chemicals for high-performance supercapacitor electrodes. *Adv. Energy Mater.*, 1, pp. 356–361.

Wei, Y. H., Li, J. J., Sun, F., Wu, J. R. and Zhao L. J. (2018). Leakage-proof phase change composites supported by biomass carbon aerogels from succulents. *Green Chem.*, 20, pp. 1858–1865.

Yeo, T. H. C., Tan, I. A. W. and Abdullah, M. O. (2012). Development of adsorption air-conditioning technology using modified activated carbon: A review. *Renew. Sust. Energy Rev.*, 16, pp. 3355–3363.

Yu, A. P., Chabot, V. and Zhang, J. J. (2013). Applications of electrochemical supercapacitors. *Electrochemical Supercapacitors for Energy Storage and Delivery: Fundamentals and Applications* (CRC Press).

Zalba, B., Marín, J. M., Cabeza, L. F. and Mehling, H. (2003). Review on thermal energy storage with phase change: Materials, heat transfer analysis and applications. *Appl. Therm. Eng.*, 23, pp. 251–283.

Zhao, L., Fan, L. Z., Zhou, M. Q., Guan, H., Qiao, S., Antonietti, M. and Titirici, M. M. (2010). Nitrogen-containing hydrothermal carbons with superior performance in supercapacitors. *Adv. Mater.*, 22, pp. 5202–5206.

Zhang, Z. and Wu, P. (2014). A facile one-pot route towards three-dimensional graphene-based microporous N-Doped carbon composites, *RSC Adv.*, 4, pp. 45619–45624.

Zheng, X., Lv, W., Tao, Y., Shao, J., Zhang, C., Liu, D., Luo, J., Wang D.-W. and Yang, Q.-H. (2014). Oriented and interlinked porous carbon nanosheets with an extraordinary capacitive performance. *Chem. Mater.*, 26, pp. 6896–6903.

Zhu, L., Lei, H. W., Wang, L., Yadavalli, G., Zhang, X. S., Wei, Y., Liu, Y. P., Yan, D., Chen, S. L. and Ahring, B. (2015). Biochar of corn stover: Microwave-assisted pyrolysis condition induced changes in surface func-tional groups and characteristics, *J. Anal. Appl. Pyrol.*, 115, pp. 149–156.

Chapter 8

Environmental Assessment of Biochar Using a Life Cycle Approach

Guillermo San Miguel

ETSI Industriales, Universidad Politécnica de Madrid, Madrid, Spain

g.sanmiguel@upm.es

Abstract

This chapter describes the environmental performance of biochar when produced using two different technologies: metal ring kiln and Missouri kiln. This performance has been calculated using Environmental Life Cycle Analysis (E-LCA) and it has been based on an extensive inventory that describes the extraction of raw materials and fabrication processes required to build the infrastructures required to carry out this transformation, and the direct gas emissions that occur during the carbonization stage. The normalized results of this analysis suggest that the impact categories most severely affected by the life cycle of the charcoal are photochemical oxidant formation, human toxicity and climate change. Most of the impact generated by the charcoal originates in the operation phase due to direct emissions generated as a result of carbonization reactions. Owing to the residual nature of the biomass considered in this investigation, the extraction of raw materials phase had a very limited contribution. The same is observed for the construction of the char manufacturing infrastructures. As an example, impact values in the climate change category were rather similar for both technologies, ranging between 3041 and 3090 kg CO_2 equivalent per ton of final product.

8.1 Introduction

Slow pyrolysis technologies have been used for centuries to transform lignocellulosic biomass (usually wood) into a carbonized solid usually referred to as charcoal or biochar. This renewable fuel has better fuel properties (higher heating value and combustion temperature, lower ignition temperatures and reduced emissions on combustion) than the biomass it originates from due to its reduced moisture content, lower volatile fraction and higher concentration of elemental and fixed carbon [Emrich, 1985; FAO, 1983; Lehmann and Joseph, 2009]. Hence, biochar may be used to replace conventional solid fuels like raw biomass or mineral coals in industrial, agricultural and domestic applications aimed at generating heat and/or power [Jirka and Tomlinson, 2014]. Other uses for charcoal reported in the literature include as a soil conditioner in agriculture and gardening applications, as an adsorbent for water and wastewater treatment, as a reducing agent in metallurgy plants, as a raw material for the manufacturing of carbon electrodes and as a medical detoxification agent [Garcia-Perez *et al.*, 2010; Lehmann and Joseph, 2009].

Owing to its renewable nature and plant origin, charcoal has the reputation to be a green and eco-friendly material. However, this assertion is not always legitimate and the sustainability of this material needs to be quantified in each case using scientific methods. One issue that needs to be considered in this analysis is that charcoal is not a unique product and its technical characteristics and environmental performance may differ widely depending on various different factors, which may include: nature of the biomass (plant species, anatomic parts, composition), biomass production method (residual biomass, purposely produced as energy crops, rainfed/irrigated systems, utilization of fertilizing/phytosanitary products, harvesting procedure and technologies), processing and transportation of biomass to the carbonization plant (size reduction, distance and transportation means), carbonization technology and conditions (temperature, heating rate, thermal efficiency, gas fraction residence times, gas recycling), charcoal processing (sorting, packing) and distribution to final user. Furthermore, a complete environmental assessment would also require quantifying the performance of the charcoal during its utilization phase (primarily direct gas emissions) and management of wastes (ash residues).

A number of publications have been produced in the last few years describing the environmental performance of charcoal produced from different biomass feedstocks [Alhashimi and Aktas, 2017; Hammond *et al.*, 2011; Ibarrola *et al.*, 2012; Iribarren *et al.*, 2012; Miller-Robbie *et al.*, 2015; Rousset *et al.*, 2011]. All of these investigations rely on the use of life cycle assessment (LCA) methodology, focusing primarily on the quantification of global warming emissions. In most cases, these investigations do not take into consideration key local environmental categories like human toxicity, eco-toxicity, photochemical oxidant formation, land acidification, land use and transformation, etc. These local impacts are primarily associated with the direct emission of volatile organic compounds during the pyrolytic stage of the carbonization process (including polyaromatics). Emission inventories for these species are not well documented in the literature, making it necessary to refer to different sources to have a more complete view of potential impacts derived from the carbonization stage of the charcoal system.

Charcoal may be produced using different technologies, which differ from each other in terms of capital and operation costs requirements, labor intensity, processing capacity, charcoal quality and direct emission. The European Biochar Foundation reports in its Guidelines for a Sustainable Production of Biochar [EBC, 2012] that most of the charcoal produced in the world is manufactured using traditional technologies with reduced carbon yields (usually below 25%) where most of the volatile fraction is released into the atmosphere with little control (often in breach of environmental regulations) (see Figure 8.1).

Missouri-type kilns are permanent structures, similar in concept but bigger in size than the rink kilns, intended primarily for the commercial production of charcoal at a larger scale. This type of system is constructed using concrete and reinforced steel to make the main carbonization chamber. The structure incorporates air inlet ports on the base to control the combustion phase, and outlet ports connected to smokestacks are used to vent the volatile products. In southern areas of Spain (Extremadura), these kilns are typically designed to accommodate the biomass loading contained in one or two 24 t trailers. Charcoal yields are similar to those achieved in steel kilns while processing times are lower (7–8 days) and loading capacities are larger (35–75 m^3), thus allowing higher throughput capacities.

In order to economize in the use of labor, optimize the utilization of infrastructures and maximize production output, Missouri and steel ring kilns are often designed to operate in batteries of two to six (or even more) units working in parallel. By doing so, plant operators may use the time

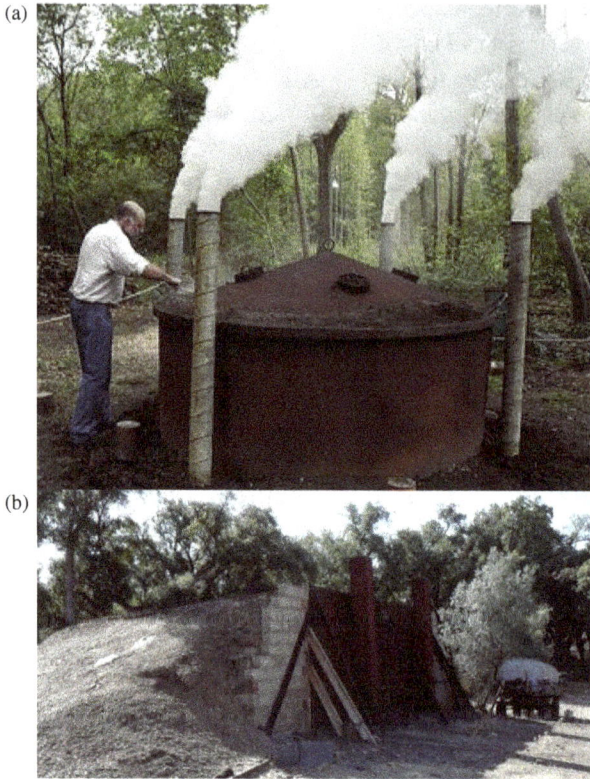

Figure 8.1:   Charcoal production technologies investigated in this chapter: (a) Metal ring kiln; (b) Missouri-type kiln.

required by one set of kilns or accomplish the pyrolysis/carbonization of one biomass loading to complete other tasks like preparation of raw materials, kiln loading and unloading, processing and packing of carbonized products, etc. This operating strategy also permits the incorporation of gas recycling systems usually consisting of a central flue and after-burner feeding into a common chimney stack. In this way, the thermal energy generated in the after-burner system during the pyrolytic stage of one set of kilns may be used by the other set of kilns to meet the high energy demands occurring during the dehydration phase.

The objective of this chapter is to describe the application of LCA methodology to quantify the potential environmental performance of charcoal. The investigation evaluates the effect of using two carbonization technologies (steel ring kiln and Missouri kiln) and assumes that the biomass employed is of residual origin. The analysis incorporates an extensive inventory of gas emissions that takes into consideration five global and local impact categories, including climate change, terrestrial acidification, human toxicity, photochemical oxidant formation and natural land transformation.

8.2 LCA Methodology

8.2.1 *Goal and scope definition*

The goal of this investigation is to quantify the potential environmental impacts associated with biochar produced using different technologies: steel ring kiln and Missouri-type kiln. The analysis has been carried out according to standard methodology ISO 14040-14044:2006 and following a cradle-to-gate approach.

As illustrated in Figure 8.2, the scope of the charcoal system investigated in this work includes four stages:

1. Collection and transport of biomass feedstocks to carbonization plant.
2. Extraction of raw materials employed to build the carbonization kilns, construction of the carbonization plants and their components.
3. Operation of carbonization plants, including gas emissions and generation of wastes and by-products.
4. Processing of final charcoal, bagging and storage.

The distribution of the charcoal to its final user and its combustion for heat generation are out of the scope of this investigation. It has been assumed that the biomass employed to make the biochar has a residual origin (e.g. clearing, pruning or cleaning of mature forests). Hence, the environmental impact associated with the generation of this feedstock has been assumed to be negligible.

Table 8.1 illustrates the technical characteristics of the system scenarios under consideration. Average transport distances of the biomass to the carbonization plant have been assumed to be 5 km for the metal ring kiln scenario and 25 km for the Missouri kiln scenario, with longer

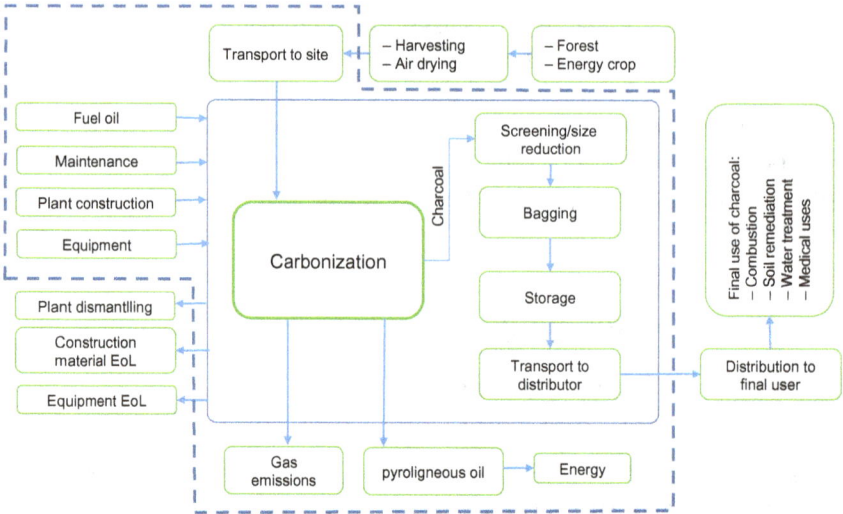

Figure 8.2:   Flowchart describing the characteristics and scope of the charcoal system.

distances attributed to systems with higher processing capacity. The smaller plant (ring kiln) has been assumed to use animal force for the transportation of the biomass feedstocks, while the larger plant (Missouri-type) has been assumed to use mechanical means (EURO 5 lorry 16–32 t). Processing capacity has been calculated considering the volume of each kiln, the apparent density of the biomass and the number of cycles per year. Fewer cycles have been assumed for the smaller ring kiln than for the larger capacity Missouri technology, taking into consideration the duration of the cycles (between 8–15 days, including dehydration, carbonization and cooling phases) and the extent of the charcoal production season (4–8 months/year, which is adapted to the availability of biomass resources and product demand).

Charcoal production yields have been assumed to be similar in both technology scenarios (22.5%) based on information provided directly by charcoal manufacturers. Bagging of the charcoal has been assumed to be done using 10 kg capacity double layer kraft paper bags, each weighing 175 g. Bagging is done manually in the small capacity system (ring kilns) and using an electric feeder consuming 1.0 Wh/kg of charcoal in the Missouri kiln. Foreground inventory data have been provided by PIROECO, a Spanish engineering and bioenergy company based in

Table 8.1: Technical and inventory data for charcoal production kilns.

	Metal ring	Missouri kiln
Apparent wood density[a] (t/m^3)	0.65	
Moisture content[a] (%)	20	
Transport to kiln (km)	5	25
Transport means	Animal force	Mechanical force
Kiln volumetric capacity (m^3)	6.5	75
Cycles/year[b]	12	18
Wood processing capacity[a] (t/cycle)	4.23	48.8
Wood processing capacity[a] (t/yr)	50.7	877.5
Product yield (wt.%) Charcoal	22.5	22.5
Condensate oils	0	0
Gases (by dif.)	77.5	77.5
Charcoal production capacity[a] (t/cycle)	0.761	8.78
Charcoal production capacity[a] (t/yr)	9.13	158.0
Kiln lifetime expectancy (yr)	5	8
Land occupation + transf.[c] (m^2/kiln)	50	500
Previous land use	Forest	

Notes: [a]Fresh wood basis.
[b]Ring kiln = 2 cycles/month and 4 month/year; Missouri-type = 3 cycles/month and 6 month/year.
[c]Per kiln, including space for charcoal storage and processing.

Malaga with extensive experience in the field. Background inventory data have been obtained and adapted from Ecoinvent 3.2.

Regarding the materials employed for the construction of the carbonization kilns, the following elements were considered:

1. **Steel ring kilns:** 290 kg of low-alloyed steel for the construction of the main structure (metal rings and cover); 46 kg of stainless steel for smoke outlets; 50 kg metal working; 15 m gas welding.
2. **Missouri kiln:** Concrete (20 m^3), steel (2,100 kg, including reinforcing steel for the main structure, low alloyed steel for doors and door frames, and stainless steel for pipes), metal working (500 kg), clay flue pipes (300 kg), gas welding (15 m), machine operation for construction (60 h) and wood construction for storage (50 m^2).

Five environmental categories have been considered in this investigation: (a) climate change, (b) terrestrial acidification, (c) human toxicity, (d) photochemical oxidant formation and (e) natural land transformation. The first one is a global impact category, while the other four represent local impact categories.

8.2.2 *Inventory of air emissions*

The gas emissions inventory is one of the most critical issues in the LCA of the biochar. This is so because of the significance of the manufacturing phase in the life cycle of the biochar system (as shown in the results below) and also the large number of gas compounds emitted during the pyrolysis/carbonization process. Most of the LCAs published in the literature describing the environmental performance of biochars are based on incomplete air emissions inventories that take into consideration only a limited number of compounds (in some cases only greenhouse gas emissions like CO_2 and CH_4).

In order to cover a wider range of gas emissions, the inventory employed in this chapter combine the results published by an array of reputable bibliographic sources for the carbonization of biomass in conventional batch reaction systems. As illustrated in Table 8.2, this inventory covers the emission of gases (CO_2, CO, CH_4, C_2H_6, C_2H_4, C_3H_6, NO, NO_2) [Bertschi et al., 2003], volatile organic compounds (VOC) (including acetic acid, formic acid, formaldehyde, methanol, phenol) [Bertschi et al., 2003], total solid particulates (TSP) [Sparrevik et al., 2015] and total polyaromatic hydrocarbons (PAH) [USEPA, 1995]. Emission values for the 16 priority PAHs were calculated based on the concentrations reported by [Mara dos Santos Barbosa et al., 2006]. Due to the comparable product yields and thermal efficiencies reported in the literature for metal ring and Missouri kilns, the same emission values were applied to both biochar systems.

Life cycle impact assessment methods ReCiPe Europe H (Midpoint and Endpoint) v1.13 were used to calculate potential impacts and damage on selected environmental categories. SimaPro v8.4 and MS Excel were used to build the models and perform calculations. The system has been analyzed using an attributional approach and adopting as functional unit 1 ton of biochar, packed and ready for distribution to final users. Assuming a standard lower heating value (LHV) for biochar of 23 MJ/kg, this mass-based functional unit may be translated into an equivalent energy-based functional unit of 23 GJ.

Table 8.2: Air emission factors (g/kg of biochar) for conventional biomass carbonization systems, as extracted from various sources.

Component	Emission factor (g/kg charcoal)
CO_2	1935 [Bertschi *et al.*, 2003]
CO	346 [Bertschi *et al.*, 2003]
CH_4	47.7 [Bertschi *et al.*, 2003]
C_2H_6	12.2 [Bertschi *et al.*, 2003]
C_2H_4	4.66 [Bertschi *et al.*, 2003]
C_3H_6	5.35 [Bertschi *et al.*, 2003]
Acetic acid	31.9 [Bertschi *et al.*, 2003]
Formic acid	1.62 [Bertschi *et al.*, 2003]
Formaldehyde	3.8 [Bertschi *et al.*, 2003]
Methanol	43.8 [Bertschi *et al.*, 2003]
Phenol	9.83 [Bertschi *et al.*, 2003]
NO	4.11 [Bertschi *et al.*, 2003]
NO_2	0.45 [Bertschi *et al.*, 2003]
TSP	8.0 [Sparrevik *et al.*, 2015]
Total PAH	0.0047 [USEPA, 1995]

PAH distribution	Particle bound	Gas phase	Total [Barbosa *et al.*, 2006]
Naphthalene	0	0.00197005	0.001970049
Acenaphthylene	1.79095E-05	0.00044774	0.000465648
Acenaphthene	2.86553E-06	0.00010029	0.000103159
Fluorene	2.14914E-05	0.00053729	0.000558778
Phenanthrene	0.000109248	0.00077011	0.000879358
Anthracene	3.04462E-05	0.000197	0.000227451
Fluoranthene	6.80562E-05	9.313E-05	0.000161186
Pyrene	5.01467E-05	7.8802E-05	0.000128949
Benz[a]anthracene	3.761E-05	5.7311E-06	4.33411E-05
Chrysene	3.22372E-05	5.7311E-06	3.79682E-05
Benzo[b]fluoranthene	2.14914E-05	4.6565E-06	2.61479E-05
Benzo[k]fluoranthene	2.14914E-05	2.5073E-06	2.39988E-05
Benzo[a]pyrene	2.86553E-05	2.6864E-06	3.13417E-05
Indeno[1,2,3-cd]pyrene	2.14914E-05	2.1491E-06	2.36406E-05
Benzo[g,h,i]perylene	2.86553E-06	1.791E-07	3.04462E-06
Dibenz[a,h]anthracene	1.43276E-05	1.6119E-06	1.59395E-05

8.3 Results and Discussion

Table 8.3 illustrates the characterized impacts of charcoal produced using
the rink kiln and the Missouri kiln technologies. Regarding carbon footprint, both systems generate similar values (3041–3090 kg CO_2/ton) due
to comparable charcoal yields, energy efficiencies and direct gas emissions during the manufacturing stage.

Impact values in human toxicity and photochemical oxidant formation categories are very similar in both technologies. This is due to the fact
that these impacts are associated with direct emissions occurring during
the charcoal manufacturing process, and these emissions are very similar
due to comparable thermal efficiencies and product yields. The higher
impact values (8.7% higher) produced by the Missouri kiln in the terrestrial acidification category may be associated with the raw materials
required to build the kiln (primarily concrete and steel) and also with the
transportation of biomass to the charcoal plant (which has been assumed
to be by internal combustion vehicle in the Missouri kiln scenario and by
animal force in the ring kiln scenario) and with the packaging system
(which has been assumed to be done using electrical equipment in the
Missouri kiln and by hand in the ring kiln scenario).

Figure 8.3 illustrates the contribution of different life cycle stages to
the impacts generated by the charcoal on selected environmental categories. The results suggest that impact values on climate change (98.6%
for ring kiln and 97.0% for Missouri kiln), terrestrial acidification (91.7%
for ring kiln and 84.4% for Missouri kiln), human toxicity (92.6% for ring
kiln and 93.1% for Missouri kiln) and photochemical oxidant formation (99.7% for ring kiln and 99.3% for Missouri kiln) are attributable
primarily to direct emissions during the carbon manufacturing phase of
the process. In contrast, impact values in the natural land transformation

Table 8.3: Characterized impacts for 1 ton of biochar produced using different
technologies.

Impact category	Units	Steel ring kiln	Missouri kiln
Climate change	kg CO_2 eq	3041	3090
Terrestrial acidification	kg SO_2 eq	2.784	3.026
Human toxicity	kg 1,4-DB eq	489.6	486.7
Photochemical oxidant formation	kg NMVOC	60.6	60.8
Natural land transformation	m^2	0.008	0.029

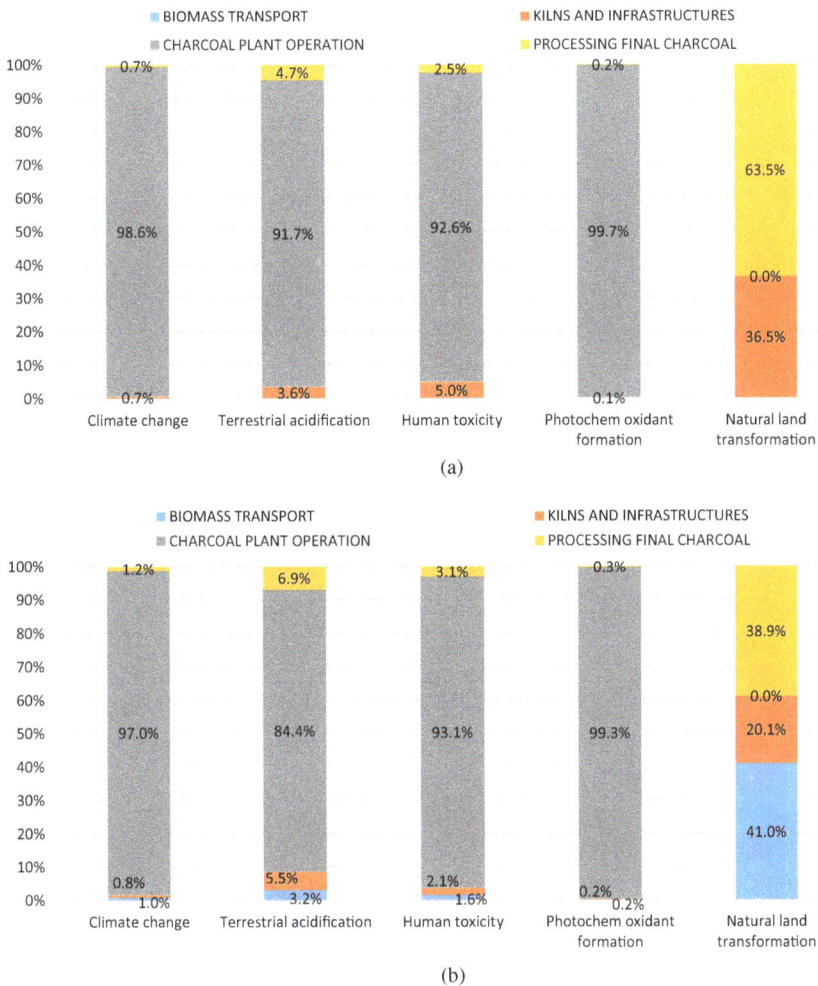

Figure 8.3: Contribution of life cycle stages to different impact categories in the ring kiln (a) and the Missouri kiln (b).

category are associated with other issues like the transportation of the biomass resources to the carbonization plant (only in the Missouri kiln scenario, which has been assumed to take place using internal combustion vehicles), the raw materials employed in the construction of the kiln infrastructures and the processing and packing of the final charcoal (which is done using electric apparatus in the Missouri kiln scenario and by hand in the ring kiln).

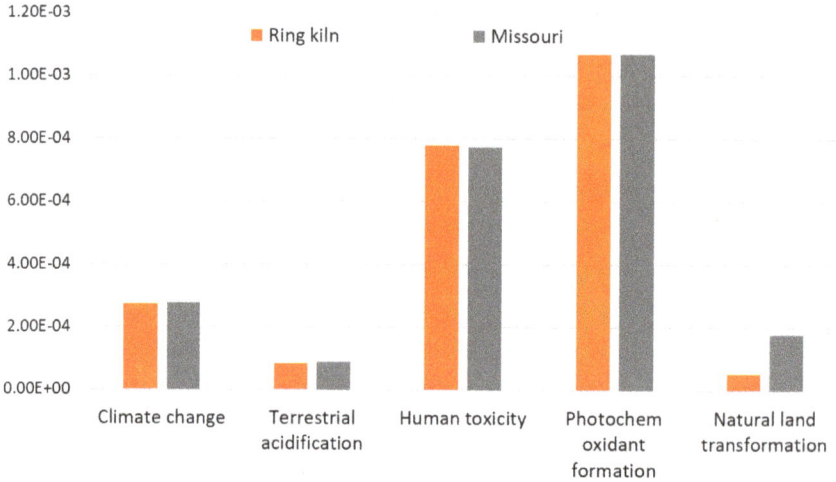

Figure 8.4:   Normalized impact values of charcoal produced using steel ring kiln and Missouri kiln technologies.

Figure 8.4 shows the normalized impacts calculated for the charcoal when produced using different technologies, as calculated using as a reference the factors proposed by the hierarchist version of ReCiPe Midpoint (H) v1.13 (Europe). The results evidence that the impact categories most severely affected by the biochar system are photochemical oxidant formation, human toxicity and climate change. These three impact categories are primarily associated with direct emissions generated during the carbonization stage. As observed in the characterized results, the differences in the environmental performance of the two technology scenarios are very limited for these four categories.

In contrast, the significance of the impacts generated by the charcoal systems (both ring kiln and Missouri kiln) on the terrestrial acidification and the natural land transformation categories is much lower than that observed for the other three categories. It had been discussed above (Figure 8.3) that potential impacts on the natural land transformation category are attributable primarily to direct natural land requirements during the construction of the carbonization plant and also to indirect land transformation requirements associated with different aspects of the charcoal system (raw materials employed in the fabrication of the kilns, transportation of the biomass, processing and packing of charcoal).

The single point impact determinations shown in Figure 8.5 evidence that the environmental performance of the two technologies investigated

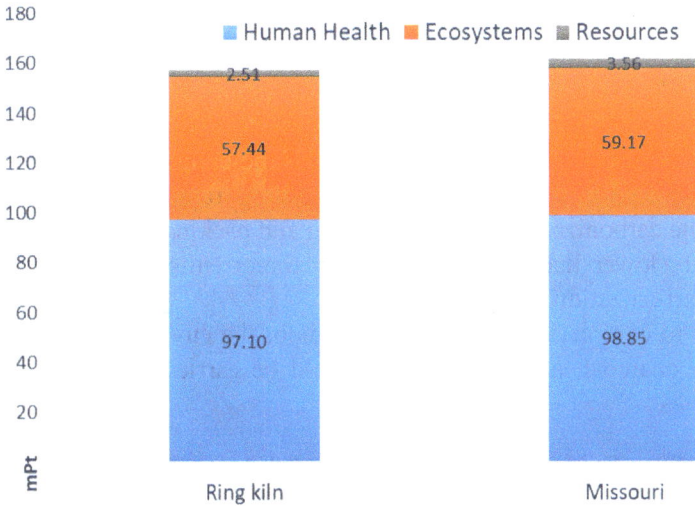

Figure 8.5:   Single point impact factor indicators for biochar produced using steel ring kiln and Missouri kiln technologies.

was very similar to each other (157.1 mPt for ring kiln and 161.6 mPt for Missouri kiln). These aggregated results evidence that most of the damage generated by the charcoal systems is related to human health (61.8% for ring kiln and 61.2% for Missouri kiln) followed by damage to ecosystems (36.6% for both technologies). Damage to the use of resources is, in all cases, insignificant due to the simplicity of the biochar manufacturing technologies in terms of material and energy requirements. These results could be very different if, instead of using a biomass residue, the charcoal had been produced from a biomass purposely grown as an energy crop.

8.4   Conclusions

The carbon footprint of charcoal produced from a residual biomass feed-stock ranges between 3041 kg CO_2 eq when produced using a steel ring kiln to 3090 kg CO_2 eq when produced using a Missouri kiln. Impact values in other environmental categories are as follows: terrestrial acidification (2.784–3.026 kg SO_2 eq), human toxicity (489.6–486.7 kg 1,4-DB eq) and photochemical oxidant formation (60.6–60.8 kg NMVOC).

The normalized results evidence that the impact categories most severely affected by the charcoal systems are associated primarily with direct emissions generated during the carbonization stage (climate change, photochemical oxidant formation, human toxicity and terrestrial ecotoxicity). The significance of other potential impacts generated by other elements in the life cycle of the charcoal systems (direct land transformation generated during the construction of the carbonization plants, raw materials used in the fabrication of the kilns, transportation of the biomass to the carbonization plant, processing and packing of charcoal) is comparatively lower than that generated by direct emissions produced during the carbonization phase.

In order to complete this investigation about the environmental performance of charcoal, additional work should be carried out along the following lines:

1. Include the utilization phase in the LCA of the charcoal system. This would be essential to compare the overall environmental performance of charcoal against other fuels (including raw biomass). This assessment would also be useful to determine whether the environmental impacts generated during the transformation of the raw biomass into charcoal are compensated by the reduced emissions of the latter during the combustion phase.
2. Reduce uncertainty in the inventory of direct gas emissions during the carbonization phase by conducting a complete experimental analytical assessment covering different charcoal technologies, production conditions (in terms of temperature, residence time, air input, carbon yield, etc.) and biomass types. This assessment should quantify the emission of gases, volatile organic compounds (including PAH), particulates and any other compound susceptible of affecting the environment.
3. Expand the scope of the environmental LCA of charcoal to incorporate economic and social issues.
4. Evaluate the consequences of not considering the biomass feedstock as a residue with null environmental burdens but as an energy crop produced in different conditions.
5. Extrapolate these investigations to other charcoal manufacturing technologies and geographical locations.

Acknowledgment

Thanks are due to Antonio Quero from PIROECO (Malaga, Spain) for inventory data regarding construction and operation of charcoal technologies.

References

Alhashimi, H. A. and Aktas, C. B. (2017). Life cycle environmental and economic performance of biochar compared with activated carbon: A meta-analysis. *Resour. Conserv. Recycling*, 118, pp. 13–26.

Bertschi, I., Yokelson, R. J., Ward, D. E., Christian, T. J. and Hao, W. M. (2003). Trace gas emissions from the production and use of domestic biofuels in Zambia measured by open-path Fourier transform infrared spectroscopy. *J. Geophys. Res.*, 108, pp. 81–91.

EBC (2012). Guidelines for a sustainable production of biochar. European Biochar Foundation (EBC), Arbaz, Switzerland. http://www.europeanbiochar.org/en/download. Version 6.2E of 04 February 2016, DOI: 10.13140/RG.2.1.4658.7043.

Emrich, W. (1985). Handbook of charcoal making, the traditional and industrial methods, Solar Energy R&D in the Europea Community, *Series E: Energy from Biomass*, Volume 7. Reidel Publishing Company.

FAO (1983). Simple technologies for charcoal making, FAO forestry paper 41 ed.

Garcia-Perez, M., Lewis, T. and Kruger, C. E. (2010). *Methods for Producing Biochar and Advanced Biofuels in Washington State, Part 1: Literature Review of Pyrolysis Reactors*, Ecology Publication Number 11-07-017.

Hammond, J., Shackley, S., Sohi, S. and Brownsort, P. (2011). Prospective life cycle carbon abatement for pyrolysis biochar systems in the UK. *Energy Policy*, 5, pp. 2646–2655.

Ibarrola, R., Shackley S. and Hammond, J. (2012). Pyrolysis biochar systems for recovering biodegradable materials: A life cycle carbon assessment. *Waste Manage*, 5, pp. 859–868.

Iribarren, D., Peters, J. F. and Dufour, J. (2012). Life cycle assessment of transportation fuels from biomass pyrolysis. *Fuel*, 97, pp. 812–821.

Jirka, S. and Tomlinson, T. (2014). *2013 State of the Biochar Industry*. A Survey of Commercial Activity in the Biochar Field. International Biochar Initiative. https://biochar-international.org/state-of-the-biochar-industry-2013/.

Lehmann, J. and Joseph, S. (2009). *Biochar for Environmental Management. Science and Technology*, Earthscan Publications Ltd., London (UK). ISBN: 184407658X.

Mara dos Santos Barbosa, J., Ré-Poppi, N. and Santiago-Silva, M. (2006). Polycyclic aromatic hydrocarbons from wood pyrolyis in charcoal production furnaces. *Environ. Res.*, 3, pp. 304–311.

Miller-Robbie, L., Ulrich, B. A., Ramey, D. F., Spencer, K. S., Herzog, S. P., Cath, T. Y., Stokes, J. R. and Higgins, C. P. (2015). Life cycle energy and greenhouse gas assessment of the co-production of biosolids and biochar for land application. *J. Clean. Prod.*, 91, pp. 118–127.

Rousset, P., Caldeira-Pires, A., Sablowski, A. and Rodrigues, T. (2011). LCA of eucalyptus wood charcoal briquettes. *J. Clean. Prod.*, 14, pp. 1647–1653.

Sparrevik, M., Adam, C., Martinsen, V., Jubaedah and Cornelissen, G. (2015). Emissions of gases and particles from charcoal/biochar production in rural areas using medium-sized traditional and improved "retort" kilns. *Biomass Bioenergy*, 72, pp. 65–73.

USEPA (1995). Emission factor documentation for AP-42, Section 10.7 charcoal, from *AP-42: Compilation of Air Emission Factors*, United States Environmental Protection Agency (USEPA), Chapter 10, Wood Products Industry.

Index

A

above-ground biomass, 94
absorptive partitioning, 144
acidic peat, 94
acidification, 195
activated carbons, 11, 132
adsorption, 10
adsorption in aqueous phase, 12
adsorption of dyes, 12
adsorption of H_2S, 14
adsorption of several heavy metals, 12
adsorption processes, 131
adsorptive heat storage, 168
adsorptive storage of CH_4, 136
agrochemicals, 90
agromining, 121, 123
air and water storage, 86
air-pollution control, 148
alkali chemicals, 10
alkaline, 94
alterations of microbial biomass, 95
aromatic polymers, 149
availability of nitrogen, 93
available metal content, 112–113

B

bacterial diversity, 96
bamboo, 70–71, 73, 75

BC high alkalinity, 93
bedding plant species, 88
beneficial plant growth, 88
benefits, 85
bioavailability, 113–114
biochar, 2, 36, 85, 87–89, 97–98, 100
biochar-based carbon materials, 167
biochar-based solid acids, 16
biochar reuse, 100
biochar-supported catalysts, 17
biochar supported metals or metal oxides, 18
biochar-supported Ni catalysts, 18
biodiesel production, 148, 150
biodiesel production processes, 17
biogas upgrading, 14, 131
biomass, 2
biomass hydrolysis and dehydration, 17
bio-oil, 36
Bortrytis cinerea and Leveillula taurica, 95
Brønsted acid sites, 148
bubbling fluidized beds, 45
Büchner funnel, 64, 66
bulk density, 65, 67, 110

C

^{13}C Nuclear Magnetic Resonance, 49

$CaCO_3$-rich biochar, 151

capacity fading, 20

carbohydrate polymers, 149

carbon-based sorbents, 131

carbon footprint, 192, 195

carbonized and KOH-activated chitosan, 168

carbon-rich coproduct, 87

carbon-rich materials, 166

carbon sequestration, 3, 5

carbon sequestration potential, 49

carboxylic acid groups, 142

catalysis, 10

catalyst supports, 16

catalytic activity, 148

catalytic or electrocatalytic transformations, 131

catalytic reduction of NO_x, 151

cation exchange capacity (CEC), 88, 108

cation–π interactions, 145

cellulose, 40

char, 36

charcoal, 87

charge-assisted H-bond, 146

chemical activation, 10, 170

chemical and physical interactions, 142

chemical functionalities in surface, 11

circulating fluidized bed, 45

climate change, 2, 6, 187, 194

CO_2 adsorption capacity, 135

CO_2 capture from flue gas, 133

CO_2 capture in postcombustion, 13

CO_2 emissions, 5

CO_2 uptake at low CO_2 partial pressures, 13

complexation, 146

condensation and polymerization reactions, 41

contaminant and nutrient sorption, 91

contaminated soil, 9, 105

cost-effective, 86

Coulombic efficiency, 19

cracking reactions, 41

crop production sustainability, 90

crop waste, 89

cucumber, 94

cyclic voltammetry, 171

D

2D crystalline carbons, 153

decomposition of raw biomass, 97

depolymerization, 41

diffusion of a sorbate, 140

diffusion rate, 13

direct catalysts, 16

dispersed metal nanoparticles, 148

disposal of waste materials, 87

drum and rotary kiln pyrolysis units, 46

dry biomass, 35–36

E

electrical conductivity (EC), 54, 108, 173

electrocatalysis, 153

electrochemical capacitors, 25

electrochemical energy storage, 19

electrode materials for supercapacitors, 165

electronic and hydrophilic properties, 172

electrostatic interactions, 145

emissions of CO_2, 2

endothermic nature, 42

energy and cost efficiency, 14

energy density, 25

energy storage, 10, 166

engineered biochars, 48

engineered hard carbons, 21

entrained flow, 43

environmental assessment, 183–184

environmental impact, 86

equivalence ratio, 39

esterification of glycerol, 150
exothermic, 43

F
fast adsorption kinetic rate, 13
fast discharge rates, 25
fast pyrolysis, 36
feedstock, 89
feedstock and pyrolysis conditions, 91
feedstock selection, 91
fertilizers may contain different
    amounts of heavy metals, 99
field capacity (FC), 62, 68, 75, 77
Fischer–Tropsch synthesis, 151
fixed-C content, 51
fluidization gas velocity, 45
fluidized bed, 43
foliar mite, 95
foliar P concentration, 93
food security, 90
Freundlich isotherm, 139
fruit weight, 94
fruit yields, 94
functional groups, 53
functionalized biochars, 16, 150
fungal foliar pathogens, 95
funnel, 72, 73

G
gas emissions, 184
gasification, 35–36, 39, 166
global warming, 165, 185
glucose-derived hydrochar, 168
greenhouse gas emissions, 190
greenhouse gases, 2
green waste, 88
growing media, 5–7, 86–87, 100

H
5-(hydroxylmethyl)furfural, 149
H:C ratio, 51
heat carrier, 44
heating rate, 36

heat of adsorption, 13
heat storage, 165
heavy metal, 9, 105–106, 112
hemicellulose, 40
heteroatom-doped systems, 153
heteroatoms, 11, 170
heterogeneous catalysis, 16
hierarchical porosity, 10
high energy density, 19
highly dispersed metal nanoparticles,
    18
high selectivity towards CO_2 over N_2,
    13
horticultural crops, 85–86
horticultural production, 97
horticulture application, 90
horticulture production, 99
hybrid supercapacitors, 175
hydrochars, 8, 165
hydrogen bonds, 149
hydrogen evolution reactions (HER),
    153
hydrogen purification, 14
hydrophobic interactions, 143
hydrophobicity, 91, 149
hydroponic production, 86
hydrothermal carbonization (HTC),
    4, 132, 166
hydrothermal conversion, 11
hydrothermal processes, 40
hydroxyl and carboxylic groups, 12
hyper accumulator plants, 107
hyperaccumulators, 119

I
impact on CO_2 emissions, 97
infiltration, 65, 66
inherent inorganic compounds, 17
inorganic fertilizers, 97
intensive crop production, 97
intercalating and accommodating
    Na-ions, 21
intermediate pyrolysis, 38

intraparticle diffusion models, 140
ion exchange, 147
ionic diffusion, 25
ionic radius, 21
ionizable functional groups, 142
iron-nanoparticle-based catalysts,
    151
irrigation, 90
ISO 14040-14044:2006, 187

K
K demanding crops, 92
$KHCO_3$-activations, 170
KOH-activated biochars, 151
KOH-chemical activation, 170

L
land acidification, 185
land and soil degradation, 97
Langmuir isotherm, 138
leaf nutrient concentration, 94
life cycle assessment (LCA), 185,
    187
lignin, 40
lignocellulosic biomass, 38, 149
liming effect, 107, 109, 112
lithium-ion batteries, 19
lithium-sulfur batteries, 19
long cycling life, 25
low-cost materials, 87
low temperature, 94

M
macrobiopolymers, 41
mass transport of ions, 172
matric potential, 68
mesopores, 91
mesoporosity, 11
metal, 113
metal-free catalyst, 16
metal mining, 105

metal oxide particles, 148
metal ring kiln, 186–187
methanation, 151
methane and nitrous oxide, 98
micronutrients, 93
micropores, 11
microporous–mesoporous structures,
    172
mine land reclamation, 105
mineral content, 149
mining soils, 112
Missouri kiln, 187, 189
Missouri-type kiln, 185–187
mobility, 114
moisture, 63, 66–70, 75

N
narrow micropores, 52
N bioavailability, 99
N-doped carbons, 133, 153, 172
N fixation, 93
nitrates and phosphates, 98
nitrification rate, 98
nitrogen, 93
N, P and K fertilizers, 98
N,S-doped mesoporous carbon, 12
nursery plant production, 88
nutrient cycles, 93
nutrient leaching, 98
nutrient storage, 100

O
ohmic losses, 173
operating conditions, 48
organic amendments, 116
organic compounds, 93
organic materials, 86, 99
ornamental plant, 88
oxygenated groups, 166
oxygen-containing functionalities, 12,
    148

oxygen evolution reaction (OER), 153
oxygen reduction reaction (ORR), 153
ozonation of gaseous ammonia, 151

P

π–π interactions, 145
P and K fertilizers, 92
paraffin compositions, 168
pathogenic organism, 95
peak temperature, 51
peat, 86
peatlands, 6
Permanent Wilting Point (PWP), 62,
    65, 68–69, 75, 77
pesticide residues, 96
pesticides, 90, 97
pH, 108, 112–113
phase change materials, 167
phenolic functional groups, 166
phenoxyacetic acids, 96
pH of biochar, 54
photochemical oxidant formation,
    185, 187, 192, 194–195
physical activation, 10, 170
physically activated, 132
physical properties, 94
phytoextraction, 105, 107, 117–120
phytomining, 120–121
phytotoxicity, 106
plant available water (PAW), 61–62,
    67–71, 77
plant-derived biochars, 89
plant diseases, 95
plant growth, 88
plant nutrient values, 92
plant production, 87
platform chemicals, 149
polarization of the C–N bonds, 153
pollutant species, 136
polyaromatic hydrocarbons (PAH),
    90, 190

polyaromatics, 185
pore, 63–65, 70–71, 77
pore-filling mechanism, 144
pore size distribution, 169
pore widening, 11
porosity, 63, 70–71, 91
porosity development, 52
potential stability of biochar, 49
power density, 25
pressure, 52
pressure plates, 64, 66, 72–76
pressure swing adsorption, 14, 136
primary pyrolysis reactions, 41
promote plant productivity, 92
promotion of mycorrhizal fungi, 92
promotion of plant growth, 87
pseudo-capacitance, 25, 170
pseudo-first order kinetics, 140
psychrometer, 65
psychrometry, 66, 73–74
pyrolysis, 2, 35, 36, 166
pyrolysis oil, 36
pyrolysis vapors, 18
pyrolytic biochars, 165

R

reactions, 41
recalcitrance index, 51
redox reactions, 142
reduction media, 18
refinery processes, 148
regeneration and reuse, 139
removal of aqueous pollutants, 131
removal of organic pollutants, 12
resistance to mass transfer, 13
Richard chambers, 64, 66
rooting media, 90
roots, 94
root system development, 86
rotating cone reactors, 43
ruthenium particles, 151

S
salt hydrates, 167
salt solutions, 168
salt templating, 11
screw or auger reactors, 45
secondary pyrolysis reactions, 41
secondary vapor cracking reactions, 38
selective adsorption of CO_2 over CH_4, 14
separation of CO_2, 131
sewage sludge, 89
single-step production processes, 11
slow pyrolysis, 38
sodium-ion batteries, 19
soil biota, 115
soilborne pathogens, 95
soilless cultivation, 86
soilless growing media, 86
soilless production, 86
soilless substrates, 98
soil microorganisms, 115, 117
soil porosity, 111
soil quality, 2, 115
soil structure, 111
solid-electrolyte interphase, 21
solution pH, 142
sorption capacity, 99, 137
sorption kinetics, 137
sorption of metal ions, 141
sorption on biochar surface, 99
specific capacitance, 169
specific capacity, 19
specific surface area, 13, 51, 169
steel ring kiln, 185, 187, 189
storage of charges, 169
substrate materials, 86
substrate pH, 88
sulfonated biochar, 149
sulfonated hydrochars, 150
supercapacitors, 19, 172
support for a PCM, 168

support plant growth, 86
surface acidic functional groups, 53
surface area, 91, 108
surface sulfonation, 16
sustainable water management, 99
syngas, 39

T
tailor-made functional materials, 10
tar cracking and reforming, 18
target organisms, 96
Technosols, 106
temperature swing adsorption, 133
tension table, 64, 66, 73–76
terrestrial acidification, 187, 192
textural properties, 170
thermal stability, 6
thermochemical conversion, 35–36
tomato, 89, 94
torrefaction, 2, 36, 38
total porosity, 91
toxic compounds, 90
toxicity, 185, 187, 192, 194–195
transesterification catalysts, 151
transportation of ions, 25

U
ultramicropores, 11, 171
ultramicropore volume, 133
University of Edinburgh's C stability tool, 51

V
vacuum swing adsorption, 133
van der Waals interactions, 144
Van't Hoff equation, 138
vapor residence times, 42
vitamin C, 94
volatile organic compounds (VOC), 185, 190
volume of ultra-micropores, 13

W
water consumption, 99
water holding, 76
water holding capacity (WHC),
   61–63, 65–66, 69–73, 75–77, 108

watering, 98
water retention, 110
water uptake, 91
win–win–win solution, 87

www.ingramcontent.com/pod-product-compliance
Lightning Source LLC
Chambersburg PA
CBHW050600190326
41458CB00007B/2118